DISCUSSIONS

ON

TECHNICAL EDUCATION,

AT THE WASHINGTON MEETING OF THE

AMERICAN INSTITUTE OF MINING ENGINEERS,

FEBRUARY 22d and 23d, 1876,

AND

AT A JOINT MEETING

OF THE

AMERICAN SOCIETY OF CIVIL ENGINEERS,

AND THE

AMERICAN INSTITUTE OF MINING ENGINEERS,

AT PHILADELPHIA, ON JUNE 19th and 20th, 1876.

NEW YORK:
PUBLISHED BY THE AMERICAN SOCIETY OF
CIVIL ENGINEERS.
1876.

PHILADELPHIA.
SHERMAN & CO., PRINTERS.

INTRODUCTION.

At a time when the minds of engineers were ripe for the discussion of Technical Education, it needed but a presentation of the case before some prominent association of engineers, to call out a full and free expression of opinion regarding the defects of existing systems and the principles and methods of improvement.

At the Washington meeting of the American Institute of Mining Engineers, in February last, the President's opening address presented the subject in certain aspects, which had the effect of stimulating discussion from the "practical" side and from the scholastic side of the profession. An interesting debate sprang up, but, as the prescribed business of the meeting did not permit time nor preparation for the suitable consideration of the subject, the Council of the Institute, upon motion of Mr. Eckley B. Coxe, appointed a committee to confer with the American Society of Civil Engineers, for the purpose of arranging a joint discussion on Technical Education, by members of both societies, during the period of their summer meetings in Philadelphia. The Board of Direction of the American Society of Civil Engineers at once appointed a similar committee.

JOINT COMMITTEE.

Henry R. Worthington, Prest., Member Am. Soc. C. E.
Thomas C. Clarke, Member Am. Soc. C. E.
Charles Bender, Member Am. Soc. C. E. and Inst. M. E.
Eckley B. Coxe, Member Inst. M. E.
Rossiter W. Raymond, Member Inst. M. E.
Alexander L. Holley, Sec., Member Inst. M. E. and Am. Soc. C. E.

The Joint Committee issued the following circular:

"At a meeting of the Joint Committee of the AMERICAN SOCIETY OF CIVIL ENGINEERS and the AMERICAN INSTITUTE OF MINING ENGINEERS on TECHNICAL EDUCATION, held in New York on March 17th, 1876, it was

'*Resolved,* That the Secretary of the Joint Committee be authorized to transmit the following request to such members of the two societies represented, and to such foreign engineers as the Committee may designate:

'You are hereby requested by the Joint Committee of the American Society of Civil Engineers and the Institute of Mining Engineers on Technical Education, to discuss this subject at a joint meeting of the two societies, to be held on June 20th, during their Philadelphia meetings, the discussion to be chiefly upon the following questions:

'I. Should a course of instruction in works precede, accompany, or follow that in the technical school?

'II. Is it practicable to organize practical schools, under the direction and discipline of experts, in engineering works?

'For the purpose of explaining more fully the nature and scope of these questions, an abstract of the remarks made on the subject at the February meeting of the Institute of Mining Engineers is herewith sent you.

'As it is believed that a number of foreign engineers, as well as many of our own experts, will wish to take part in this discussion, the time of each speaker will be limited to ten minutes, until after all on the list have spoken, but any reports or considerations on Technical Education which may be presented to the meeting in writing will be printed with the proceedings.

'Please give early notice if you will take part in the discussion.

'A. L. HOLLEY,
'Secretary.'"

This circular was sent to several prominent engineers, foreign and resident, who were not members of either society, and to a number of such members of the two societies as would probably represent the widest range of experience and opinion. Of these, nearly all expressed a desire, and the greater part an intention, to join in the discussion; of the latter, some were finally prevented, by various causes, from carrying out this intention. A few gentlemen took no notice of the invitation. More than half those applied to, took part in the debate or they furnished papers.

The Joint Committee direct that the address of the President of the Mining Engineers, and the debate thereon at Washington, which are in fact a part of the same discussion, shall be affixed to the record of the joint discussion, so that the whole matter may be presented consecutively and in order.

A. L. Holley,
Secretary of the Joint Committee.

New York, Oct. 1, 1876.

ADDRESS OF PRESIDENT A. L. HOLLEY,

BEFORE THE

AMERICAN INSTITUTE OF MINING ENGINEERS, AT WASHINGTON, FEB. 22d, 1876,

ON THE

INADEQUATE UNION OF ENGINEERING SCIENCE AND ART.

THE application of scientific methods to the investigation of natural laws and to the conduct of the useful arts which are founded upon them, is year by year mitigating the asperity and enlarging the outcome of human endeavor. More notably, perhaps, are these the facts in that system of productive and constructive arts of which engineering is the general name. In metallurgical engineering especially, within the period of our own recollection, how rapid has been the rate and how wide the scope of progress: the scientific discovery and mining of metalliferous veins; the economical separation and reduction of ores of every grade; the production and regulation of high temperatures; the varied improvements in the manufacture of iron, in saved heat and work, in uniformity and range of products; and, most important of all, the creation and the utilization, to be counted by the million tons a year, of the cheap constructive steels.

Wonderful as this range and degree of development may appear to the public eye, the close and thoughtful observer must, nevertheless, conclude that neither the profession nor the craft of engineering may congratulate themselves too complaisantly, but that they should rather acknowledge to each other the embarrassing incompleteness of the union between engineering science and art.

There is a small school of philosophers whom we may designate as original investigators, men who come close to nature, who search into first principles, and who follow that scientific and therefore fruitful method by which the relations of matter and force are discovered, classified, and brought within the reach of practice. These wonderful men do not indeed create the laws of nature, as they sometimes almost seem to, but they go up into the trembling mountain and the thick darkness and bring down the tables upon which they are written.

There is a larger class of men whom we may designate as the schoolmen; they are learned in the researches and conclusions of others, and skilled in reasoning or speculating from these or from abstract data upon the certain or probable results of physical and chemical combinations.

And there is the great army of practicians, almost infinite in its degrees of quality, ranging from the mere human mechanism by which mind lays hold of matter and force, through all the grades of practical judgment and power.

Let us first consider the matter from the "practical" man's standpoint. Every day's experience teaches him that the men who speculate, from secondhand data, upon the probable results of combinations of forces and materials, are not the men who can best make these combinations in practice, who intuitively know all the concealed pitfalls, such as friction, that trick of nature which like the thousandth part of phosphorus, alters all the conditions of use in iron; nor are they the men who can determine the completeness of these combinations, or read the record of their results, as in the character of a flame, in the feeling of a refractory mixture, in the behavior of a metal under treatment; nor are they the men who, by familiarity with objects and phenomena, are best fitted to pursue that original investigation which is the foundation of even theoretical progress. The expert who delights to call himself "practical," is honestly amazed at the attempts of experts by school graduation, who have not been graduated in works, to solve the engineering problems of the day. And from his standpoint there are numerous and conspicuous illustrations. While metallurgists are still disputing over the nature and sequence of reactions in combustion and reduction, the practical ironsmelter has felt his way from the barbarous practice of a century ago, to the vast and economical production of to-day. The attainment of powerful and sufficiently hot blast by means of waste heat, the adaptation of shape and proportion of stack to different fuels and ores, labor-saving appliances and arrangements,—all these have grown out of the constant handling, not of books, but of furnaces.

Proceeding upon a chemical knowledge little superior to that of the average schoolboy, Bessemer developed his revolutionary process. Not knowing for years that the combustion of silicon or of manganese are the chief sources of the necessary heat; ignoring the fact that not alone the reaction but the presence of manganese is a cause of soundness and malleability in steel; magnifying the hypothesis that silicon should promote soundness; instructing his licensees to avoid all irons containing above 0.02 per cent. of phosphorus; and sharing the ignorance of the whole metallurgical profession as to the sequence of reactions in the converter and the probability of changing their character, Bessemer and his followers, during the first fifteen years of their practice, nevertheless brought this difficult art, which the metallurgical schools call a chemical art, to a high degree of commercial success, and this in the absence of any metallurgical change or chemical improvement whatever, in the treatment of the metal. During all this time, there was almost no literature of the Bessemer manufacture, and no instructor save that grim sphinx the converter and the well-nigh inscrutable process. It was a hand-to-hand fight, involving mechanical details, refractory linings, celerity of operations, regularity of melting and conversion and economy of labor. With every fact written in his book, the closeted scientist could no more adequately prescribe the practical conditions of improvement, than could the student in optics specify in words and formulæ the glory of an Italian sunset.

Here is a cupola-furnace, an old and exceedingly simple device; but one may know all the laws of combustion and fluxing that are written in the cyclopedias, and yet fail to change its working at will, or fail to detect the coming change, until by long familiarity, the phenomena reveal themselves as it were instinctively. One may have learned every law of the reactions of oxides and fluxes upon a refractory material, yet until his practiced hand and eye and ear can nicely detect its physical qualities and measure the results of new ingredients and temperatures, he may wander for years in a maze of uncertainties. Notwithstanding all our previous knowledge about the inevitable combustion of carbon and oxygen in the presence of heat enough to ignite them, the Siemens-Martin process, both in its calorific and in its metallurgical aspects, was as purely unpractical as the direct utilization of sun-heat is to-day, until after years of patient observation, not chiefly by scientists but by men unacquainted with books and knowing nothing at second-hand, innumerable small increments of improvement at last produced a sufficient temperature in a durable furnace.

In the development of machinery, the same history is repeated. The proportions of parts, in fact, the modern formulæ themselves, are derived from the study of innumerable experiments. The adaptation of machinery can only be perfected by him who, as it were, enters into it, making it an incarnation of himself. This enlargement of a man's organism is most strikingly illustrated in the locomotive. Oliver Wendell Holmes has happily described this putting of his life into his "shell" boat, his every volition extending as perfectly into his oars as if his spinal cord ran down the centre of its keel, and the nerves of his arms tingling in the oar-blades. The thoughtful locomotive-driver is clothed upon, not with the mere machinery of a larger organism, but with all the attributes of a *power* superior to his own, except volition. Every faculty is stimulated and every sense exalted. An unusual sound amid the roaring exhaust and the clattering wheels tells him instantly the place and degree of danger, as would a pain in his own flesh. The consciousness of a certain jarring of the foot-plate, a chattering of a valve-stem, a halt in the exhaust, a peculiar smell of burning, a sudden pounding of the piston, an ominous wheeze of the blast, a hissing of a water-gauge—warning him respectively of a broken spring-hanger, a cutting valve, a slipped eccentric, a hot journal, the priming of the boiler, high water, low water, or failing steam—these sensations, as it were, of his outer body, become so intermingled with the sensations of his inner body, that this wheeled and fire-feeding man feels rather than perceives the varying stresses upon his mighty organism.

Mere familiarity with steam-engines is not, indeed, a *cause* of improved steam-engineering, but it is a *condition*. The mechanical laws of heat were not developed in an engine-house, yet without the mechanism which the knowledge derived *through this familiarity* has created and adapted, the study of heat would have been an ornamental rather than a useful pursuit. So in other departments. When one can feel the completion of a Bessemer "blow" without looking at the flame, or number the remaining minutes of a Martin steel charge from the bubbling of the bath, or foretell the changes in the working of a blast-furnace by watching the colors and structure of the slag, or note the carburization of steel by examining its fracture, or say what an ore will yield from its appearance and weight in the hand, or predict the lifetime of a machine by feeling its pulse; when one in any art can make a diagnosis by looking the patient in the face rather than by reading about similar cases in a book, then only may he hope to practically apply such improvements as theory may suggest, or to lead in those original investigations upon which successful theories shall be founded.

These are the conclusions of the "practical" man, and they are none the less true because they are not the whole truth. That they are too little considered by the schoolmen and the graduates of schools is also true, but happily, less conspicuously so as the years advance.

The evil consequences of this mistake develop themselves in various ways. The recent graduates of schools do not, indeed, expect immediate positions of responsibility and authority, but they often demand them after too short a term of object-teaching. Perhaps the greatest advantage of their scientific training is that they can learn from objects and phenomena faster than can the mere workman, who, although full of the elements of new and useful conclusions, lacks, if I may so say, the scientific reagent which precipitates the rubbish and leaves a clear solution of the problem. It is however true—in the iron manufacture, perhaps, especially true—that men of wide learning and of great mental dexterity, unless they have studied at least as many years in the works as they have in the school, do not successfully compete for the desirable places with the men who have come up from the ranks. Narrow, unsystematic, and fruitless of new results as his knowledge may be, he who has grown up steadily from the position even of puddler's helper, will be selected to take the manager's post in preference to him whose reputation is founded solely on the school.

Nor does this prove, as the schoolmen too often believe, that the owners and directors of metallurgical enterprises are generally unappreciative of scientific culture. It rather proves that the lowest functions, as in the case of poor humanity, must first be considered; that the conditions of maintenance and regular working, which constant familiarity with objects and phenomena alone can provide, are earliest in order. Conservation first and improvement afterwards.

Another consideration in this connection is that scientific aid appears to be more readily provided for the "practical" man than practical aid for the "scientific" man. The trained scholar can the more readily adapt himself to the situation. He should suggest many more improvements than would ever crystallize in an equally good but undisciplined mind. Yet his attempt, with mere scholastic aids, to carry these improvements out, might disorganize a whole establishment. As there must be one final authority, judgment founded on experience almost universally ranks the wider and more fruitful culture of the school. And if we ask those great masters whose experimental knowledge is as wide as their scientific culture, they will tell us, that as the inert and clumsy flywheel, that typical conservator, is more helpful to a steam-engine in the long run, than a valve-gear so highly organized that it seems to know what it ought to do, so in their own undertakings, plodding, practical economics must sit in judgment upon theory and limit the reaches of imagination.

Another evil growing out of the inadequate regard of mere schoolmen for practice, is the frequent failure of their works or their inability to complete them. Inventions and constructions, designed after a scientific method and under the light of organized facts and detailed history as laid down in books, may fail simply in default of a practical knowledge of how far the capital at hand will reach, or what the means at hand will do, or what the materials at hand will stand, or what the labor and assistance at hand can be relied on to accomplish. A vast number of facts about the operation of forces in materials are so subtle, or so incompletely disentangled from groups of phenomena, that they cannot be defined in words, nor understood if they could be formulated. But after long familiarity with the general behavior of materials under stress, a practical expert can, by a process more like instinct than reason, judge how far and in what directions he may safely push his new combinations. Thus while the unschooled practician so usually wastes his energies in unscientific methods and on impossible combinations, but generally carries into successful use his comparatively few well-founded attempts, the student merely of principles and abstract facts so usually originates the ideas upon which progress is founded, and so rarely clothes them with practical bodies. In this chasm between science and art, how much effort and treasure, and even life, are swallowed up year by year.

These are not theoretical considerations. The blast-furnace, the converter, and the open-hearth have already been referred to; let us observe some other illustrations. A bridge-builder will tell us that few structures in his department of engineering fail by reason of mistakes in calculating the strain-sheet, but that the majority of failures arise from vibrations, buckling, rapid wear of important parts, shapes that weaken the material, inequalities in the material, and similar causes which are not stated in books, which assume different aspects under every change of proportions and dimensions, and which can only be inferred by means of a long familiarity with the behavior of similar structures during varying periods of service, and with the processes by which materials and members are fabricated. The builder of a machine like a marine engine, or a locomotive, or a roll-train, or a steam-hammer, will tell us that, in designing new adaptations, after every stress that can be distinctly analyzed is provided for, mass to resist vibration, changes of shape to insure sound casting, and various modifications which cannot be formulated for the want of even approximately complete knowledge of their conditions, must still be supplied, simply by judgment founded on long observation of phenomena under similar conditions. And he will thus explain nine-tenths of the failures. Who can imagine the volume of a book, or of an author, which should adequately teach the principles of construction as affected by the chiefest of all practical considerations—the economies of the foundry, the forge, and the machine-shop? With the tools and facilities at hand, what divisions of a particular structure, what shapes and sizes and methods of joining can be made cheaply as well as strong and efficient, in all the infinite forms of mechanism? Obtaining such facts from any other source than personal practice, would be like an oarsman studying a book to know when and how in the race he must husband his power, or like a wrestler looking out in a cyclopedia the probable feints of his antagonist. The successful constructor will assure us that no possible training in the school, nor any genius in invention can build economically without such a knowledge of the shop as the athlete has of the possibilities of muscular strength and agility.

These arts have been selected as examples, not because they chiefly depend on skill, but because they so largely involve the highest formulated mathematical knowledge. How much more important, then, is practical training in those departments where physical laws are very incompletely understood and formulated. How far short of practical success will abstract science stop in sinking pneumatic piles through wrecks and boulders, in tunnelling rocks traversed by subterranean streams and beds of quicksand, in cheaply applying hoisting, ventilating and draining machinery to mines where the scene and conditions of operation are constantly shifting, in firmly founding heavy and vibrating machinery on treacherous ground, in handling and casting melted steel, in constructing refractory metallurgical vessels, in delivering bars red-hot and crooked in infinite directions to a roll train, in fabricating durable breech-loading cannon, in building boilers that shall provide for vaporization, circulation, separation, cleaning and durability, in designing enginery like the horseshoe machine to shape metals, in proportioning gas-furnaces, in submarine warfare, in aerial navigation, in machine tools, in traction engines, in scaffolding and erection, in railway running-gear, in forming artificial stone under water, in permanent way, in coal-cutting, in dredging machinery, in moulding and casting, in brick machinery, in tube-drawing, in coal-burning, in pavements? Limited or impossible as would be the progress of engineering arts in the absence of that knowledge and those methods which are imparted in schools, delay and failure would hardly be less conspicuous if the schoolmen should stay in the schools and thence attempt the application of abstract science, or expect mere workmen to apply it by hearkening to their directions.

I hope it may not seem that the dignity of abstract scientific investigation is undervalued by the utilizers of nature's powers and materials, or that any considerations of profit obscure, even in the average commercial mind, the splendor of those achievements made in the mere love of truth, with thought of neither commercial application nor pecuniary reward—achievements which distinguish such names as Faraday, Bunsen, Leverrier, Mayer, Joule, Henry, Darwin, and Tyndall. Do not their successes rather encourage us, in our lower sphere, to more persistently pursue the method of these great discoverers—the original investigation of Nature's truths? Not less literally than in the poet's fancy,

> "To him who in the love of Nature holds
> Communion with her visible forms, she speaks
> A various language."

To the skilled artisan she reveals herself as truly though not as widely as to the philosopher. In the aphorism of Goethe, "Mankind dwell in her and she in them. With all men she plays a game for love, and rejoices the more they win."

But the undervaluation of the study of objects and phenomena by schoolmen, is not the principal hindrance to the complete union of science and art. A greater obstacle is the combined misapprehension and ignorance on the part of a large class of "practical" men, of what they are pleased to call "theory," meaning by theory, something which is likely to be discordant with fact—or possibly with the interests of the craft. We can hardly complain that their objection is ill-grounded, as far as it is grounded upon the practice of theoretical men; but the world has a right to complain of their narrowness of observation, of their stolid incomprehension of the results of science, of that pride of ignorance, of that bigotry, of that positive fear of the diffusion of knowledge, which is the normal condition of those who range only within the sphere of their own practice, and to whom analysis and generalization, in their business affairs, as well as in morals and politics, are an unknown thing. It is unfortunately true that a large number of managers in metallurgical enterprises—men who are deemed indispensable, and who probably, are indispensable, in the average state of practical science, are thus not incorrectly characterized. Conscious of their power as conservators, ignorant of the elements of improvement, and not unfrequently jealous and blindly fearful for the interests of their craft, they sit triumphant on an eminence (the steady undermining of which they cannot observe), and sneer at the too frequently condescending magniloquence of recent graduates and book men. The best of this class are the workful and painstaking men who come up from the ranks—men who are plucky in emergencies and regulative of labor—men whose unconscious reasoning or intuition covers the ordinary exigencies, and who, perhaps for this very reason, never inform themselves outside of their own range of observation, nor observe in a methodical or fruitful manner.

There is also a class of practicians who do secretly and abstractly respect the labors of the scientific investigator, and are unwillingly governed, more or less, by his conclusions; but their minds are so barren of general facts and so untrained in the scientific methods of utilizing facts, and hence so distrustful of any ideas which reach beyond their own practice, that they also are impediments rather than helpers in the union of science and art.

It is often said, I am aware, that there is never any real antagonism between science and art, and that all men respect, even if they do not promote, the efforts of both scientists and practicians to forward the useful arts. What then shall we say of that phase of trades-unionism, which not only tends to repress improvement, but which often violently defeats the works of progressive thinkers and sometimes destroys their authors? Let us also observe an extreme, but not isolated case of the executive treatment of science. Long before the professional career of most of us began, the Erie Railway Company commenced a series of experiments in civil and mechanical engineering, sometimes elaborate, like those of Zerah Colburn on traction, and always useful; many of them incorporated with and improving the practice of the road for a quarter of a century. The voluminous drawings and records of this experimental practice, always preserved by the engineers of the road, were just beginning to be remembered by young and inquiring engineers, as a mine of professional information, when it was discovered—you will hardly believe me—that the engineer of the Erie road having been turned out, the whole of this priceless accumulation of reports and drawings was dragged off by the cartload to a paper mill, and destroyed.

by James Fisk, Jr. In reviewing the railway history of the country, many of you will remember that this act of vandalism has been by no means the worst blow which engineering has received from so-called "practical" men.

I have referred to these exceptional cases, merely to correct a modern idea that engineering progress, especially by scientific methods, is as yet, the creature of popular favor. It is refreshing to turn from such considerations to the still exceptional but happily growing appreciation and helpful respect of practitioners and scientists for each other, as sometimes exemplified in the various departments of mining engineering. When we see recent graduates patiently leading the untrained, confused, but determined mind of the workman, painfully wrestling with hard names and occult processes, into methodical habits of thought and the rudiments of organized knowledge; when we see the grimy workman, not standing aloof for fear of his craft or of his trades-master, but dragging the recent graduate into mines and furnaces, and patiently teaching him how to recognize that matter, in mass and under mighty forces, which he had heretofore contemplated in cabinet specimens, and chiefly in ideas; when we see the commercial manager of metallurgical enterprises opening his works to the graduates of schools, and giving them a chance, not only to complete their education, but, by judicious application of their efforts, to earn a living meanwhile; when we see such things, as happily we may, here and there in metallurgical works, we may assure ourselves that *one* way has been discovered to promote the union of science and art.

In the enlargement of this method of mutual respect and instruction, to a certain extent lies the solution of the problem under consideration; but it is a complex method, only actively operative under several important conditions, such as:

1. A *public opinion* among schoolmen that a course of object and phenomena study in works is to be reckoned, not as a matter of mere business sequence, but as a large and equal feature of that curriculum which is essential to a degree of professional graduation.

2. A diffusion, among the class which we have termed the "practical" class, of a real appreciation of an organized system of information and of the scientific method of making this information useful to all classes of men and noxious or unimportant to none; such a general explanation to that vast, preponderating class of workmen and of foremen and managers, who are foremen and managers simply because they have been efficient workmen, as will

ever prevent their indiscriminate and contemptuous application of the term "theory" to whatever a schoolman proposes.

3. An understanding among the owners, directors and commercial managers of engineering enterprises, that it is not a matter of favor, but a matter of as much interest to themselves as to any class, that young men of suitable ability and of suitable preliminary culture, however acquired, should have opportunity and encouragement to master the practical features of technical education in works, not as mere apprentices, but under reasonable facilities for economy of time and completeness of research.

But these conditions do not largely exist, and are only growing with general civilization. They must be hastened and magnified by some better means than merely stating the case again and again, as some of us, I confess, are too fond of doing; than perpetually repeating, in a manner more sentimental than efficient, that scientists should appreciate practice, and practicians should appreciate science, and capital should join the hands of science and practice, saying: "Bless you, my children," in the expectation that this will prove a fruitful union. Let us rather inquire if some new order of procedure in *technical education*, some revolutionary innovation, if need be, will not put the coming race of engineers on a plane which is lifted above the embarrassments from which we are slowly emerging.

1st. In order that the technical school should be in the highest degree useful, fruitful, and economical, it must instruct, not *men* of good general education, but *artisans* of good general education. The art must precede the science. The man must first feel the necessity, and know the directions of a larger knowledge, and then he will master it through and through. Mark how rapidly the more capable and ambitious of practical men advance in knowledge derivable from books, as compared with the progress of bookmen, either in books or in practice. Many men have acquired a more useful knowledge of chemistry, in the spare evenings of a year, than the average graduate has compassed during his whole course. These men realized that success was hanging on their better knowledge. Familiar with every changing look of objects and phenomena, they detected the constant play of the unknown forces which underlie them, and longed for a guide to their operation, as a mariner longs for a beacon light. This practical familiarity and judgment at once revealed the importance of scientific facts and methods, promoted their acquisition, and guided their application. Under what comparative facilities does the mere recitation-room student, or even the mere analyst of the hundred bottles, study applied chemistry? It is to these a matter of routine duty, without a soul; they are neither stimulated nor directed by a previously created want. Beginning with theoretical and abstract knowledge, is no less an inverted process in the useful arts than in the fine arts; as it would be to take a course of Ruskin within brick walls, as preparatory to opening a studio, and then climbing the mountains to square nature with the book.

Undoubtedly there may be extremes in any form of educational method. For a youth to begin the special business of technical education by any method, practical or otherwise, before he has acquired not only a common school education, but, at least such a knowledge of polite literature and general science, including of course mathematics, as would fit him to enter one of the classical colleges, should be strongly discouraged, for various reasons. It is useless to disguise the fact that the want, not of high scholarship, but of liberal and general education, is to-day the greatest of all the embarrassments which the majority of engineering experts and managers encounter. This statement cannot be deemed uncomplimentary to the class, seeing that they have risen to power despite the embarrassment. At the present day, the high-school systems founded by States and by private enterprise, bring such an education within the reach of every one; and it seems of the first importance to promote, if not almost to create, a public opinion, that liberal and general culture is as high an element of success in engineering as it is in any profession or calling.

But this is not all. Professional and business success are not, even in America, the chief end of life. All the social and political relations, and even personal happiness, are governed, not by the specialties, but by the balance of mental culture. What, then, shall we say of the policy of wealthy parents—not indeed general, but too frequent—of placing an uncultured boy in a technical school and then in works and business, without giving him one chance to acquire a general and polite culture?

Many young men display a liking, and others a marked talent, in some special direction. There is no danger that these will be crowded out of existence by the culture necessary to make a well-balanced mind; and the nearer the talent approaches genius, the less imminent will be any such danger.

The proposition then is, not that mere common school-boys shall go into works and then into technical schools, but that young men of more advanced general culture, when they do begin the business of technical education, shall apply to Nature first and to the schoolmaster afterwards.

It may be urged in favor of beginning in the technical school, rather than in the works, that mental capacity for the after acquisition and application of facts and principles is thus developed. But mental training is not the product of the technical school alone. Habits of logical thinking and power of analysis and generalization may be acquired in any school. And a positive objection to beginning with the technical school is, that it cannot stop at logical methods and sciences which are essentially abstract. It also attempts to teach about objects and phenomena, the first knowledge of which, if it is to be broad and genuine, must come from the fountain-head.

These considerations may be farther illustrated by the course of the inexpert graduate when he enters works as a matter of business or of study. We have seen that the practical man can, at least, keep the wheels running and the fires burning, and that when he is of a certain grade of ability and ambition, he will most rapidly acquire the scientific knowledge and culture which, joined to his practical judgment, make him a master. The unpracticed graduate, however, can keep neither wheels turning nor fires burning; he has not even the capacity of a conservator. Nor can he for a long time recognize, in the whirl and heat of full-sized practice, the course and movement of those forces about which his abstract knowledge may be profound. The youngest apprentices are more useful in an emergency. He must begin with the lowest manual processes, not indeed to become simply dexterous, but, as it were, to learn the alphabet of a new language. He has started in the middle of his course instead of at the beginning. He must go back before he can advance, while the practician goes straight on. The knowledge of the schoolman about physical science, however often he may have visited works and mines and engines during school excursions, is essentially abstract; it no more stimulates desire and power of practical research than the calculus creates a passion or a capacity for studying the actual work of steam in an engine, or the actual endurance of a truss in a bridge.

The disappointment of inexpert graduates at finding themselves so far from being experts, their inability ofttimes to pay for farther schooling, the necessity that they should now begin to earn money, as they had persuaded themselves they could so readily do upon graduation, discourage many from pursuing engineering, and, what is worse, send many out into practice who never do complete their technical education, but who, by the character of their work, lower the professional standard.

It can hardly be urged against the precedence of practical culture, that the student will get "out of practice" while he is in the school. He may, indeed, lose dexterity, but not the better fruits of experience. In fact, those who begin as practicians, almost instinctively keep up their intimacy with the current practice.

A most signal advantage of beginning technical education in the works is, that the mind is brought into early and intimate consideration of those great elements of success which cannot be imparted in any other way,—the management of labor and the general principles of economy in construction, maintenance, and working. *An early knowledge of these subjects moulds the whole character of subsequent education and practice.* There seems to be no corresponding advantage in beginning with the technical school. The fundamental mathematics and general information on physical science may be acquired in the preliminary school.

There is little doubt that the managers of technical schools will favor this order of study. They want to graduate, not half-educated men, but experts. They desire, of all qualifications in the student, that enthusiasm which can only spring from a well-defined want of specific knowledge.

2d. But the *order* of education is not the only desirable change. Whether before or after their course in the school, the hundreds of young men who are every year entering engineering pursuits, are wasting their time in bad methods of practical study, or, if after the school course, they are more frequently doing bad work as engineers, when they should still be only students. Hardly two engineers acquire any part of their practical knowledge in the same curriculum. They pick it up as best they may, usually in a manner that is wasteful of time or damaging to the public. While the teaching of general facts and principles and of scientific method is highly developed, there is no organized system for guiding students to direct knowledge of objects and phenomena. This statement requires two explanations: I. Apprenticeship is a school of skill in a specialty rather than a school of liberal art. It is intended for a class of men who propose to remain mere workmen, and not for the class who intend to improve and direct engineering enterprises. It imparts a degree of dexterity far beyond the requirements of the general expert, while it would hardly impart in a lifetime his required range of practical knowledge. II. A school of engineering practice, such as that of research in zoology which was established by Agassiz, would be wholly impracticable, because it could be nothing less than a vast and successful establishment for construction and operation in nearly all the departments of engineering. If such a school were not commercially successful, and if its range were not comprehensive, it would be unsuitable and inadequate.

Now, if there can be a *system* of instruction in the one school, there can be in the other. The same discipline and responsibility, the same guidance as to precedence of study, quality of evidence, and correctness of conclusion, should hold good in both cases. To say otherwise would be to say that *all* knowledge should come from unaided original research, and that every investigator should begin, not where a former investigator left off, but where he began. It therefore appears that there can be a school of practical engineering, but that it cannot be mere apprenticeship in engineering practice, nor a system of engineering construction and operation, maintained merely for the purposes of a school.

The only alternative is to establish organized schools in the various existing engineering works. At first, this idea would seem subversive of all discipline and economy, but I am assured by experts in several branches of engineering, that such would not be the case. Let us take, for example, a Bessemer works. A score of students under the discipline, as well as under the technical guidance of a master, could be distributed among its various departments, not only without detriment, but with some immediate advantage to the owner, for while receiving no pay, they would become skilful, at least as soon as the common laborers who form the usual reinforcements. Students should, of course, be expected, not to work when and in what manner they might choose, but to do good and full work during specific hours. This responsibility as workmen, would rapidly impart not only the knowledge sought in the works, but a desire for higher knowledge and culture.

These considerations are not merely theoretical. Several students at a time, subjected to no discipline, sometimes working hard, and sometimes not at all, may often be found in a Bessemer works, and I have yet to hear of their embarrassing the management in any way. The laborer has no cause for interference, as the students are not under pay, and whatever they accomplish is clear gain to the three parties concerned,—the owner, the student, and the operative. A large number of young men may be found studying in machine shops, and sometimes earning small pay, besides having opportunity to work in all departments.

The proposition is to enlarge and systematize the existing desultory study in works—to increase its usefulness to the student, and, at the same time, to make the granting of such facilities to students an object, immediately, as well as remotely, to the owners of works. To this end, the schoolmaster should be not only well read in the professional literature, but a practical expert who could take charge of the works himself, so that whilst best aiding the students, he could prevent their interference with the regular and economical operations. His functions would be, not those of an instructor, nor, to any great extent, of a clinical lecturer, but those of a disciplinarian. The students should acquire skill, in order that they might acquire judgment of skill and original knowledge of materials and forces, and the master should see that they did acquire them all. He might do some service by stated examination and current criticism and suggestion, but his chief office would be to promote honest work, and to provide opportunity for work in all departments with reference to the economy of the student's time and to the owner's interests.

It should thus appear that these somewhat radical changes in the curriculum of engineering study—first, a hand-to-hand knowledge, acquired not desultorily, but by an organized system, and afterwards the investigation of abstract and general facts and their relations, would largely economize the student's time and better the quality of his knowledge. The novice is nearly as valuable a student in works as the graduate, but he is a vastly less apt scholar in the school. My own belief, founded on the study of many typical cases, is, that this order of procedure would produce a better class of experts in little more than half the time required by the reverse order; that it would always make *experts;* that it would discourage none from finishing an engineering education which would be complete in its parts, even if insufficient time were taken to fully develop it. A well-balanced culture will naturally grow in scope and in fruitfulness.

In this connection it seems proper to say a word about the royal road to learning, which a few ill-advised students attempt to pursue I do not refer to their availing themselves of professional data and drawings on file in engineering offices, but I do refer to their asking engineers and managers to furnish them special reports on subjects regarding which their own observation would be vastly more useful to the applicants, and quite as convenient to the respondents—reports on the number and duties of workmen in each department, and the particulars of operation and relative cost, which can only be profitably investigated by a student, when not only the facts but the reasons are ferretted out by himself, rather than transmitted to the academic grove through the post office.

In conclusion, if it should appear upon larger observation, to the profession in general, as it does appear to many of its members, that this want of coalescence, ranging from indifference to antagonism between its scientific and practical branches, is a real and substantial fact, a larger effort would undoubtedly be made to change a condition so damaging to the profession and to the public. This inappreciation of one department by the other is not unnatural—neither side has taken sufficient pains to observe what the other side has done. The mere scientist instinctively believes that the achievements of the profession are so far due to the deductions of scientists that all other causes fade into insignificance; and the practician knows that just as far as animal life is from the disembodied spirit, so far is utilization of nature from the formulæ of heat, chemical affinity and mathematics itself.

The first step is to recognize the fact, and I beg engineers, especially those who from their scholastic habits, see least of the everyday embarrassments which are encountered by the executive departments of the profession, to take into account, not only the pride of class power, which the artisan feels as keenly as the scientist, but those baser elements of disunion, ranging from trades-unionism to counting-room dictation in technical affairs.

Having recognized the grave and comprehensive character of the evil, the next step should be, not I think, to attempt any violent alteration in the existing conduct of engineering by the men who are now in active service, but to change, if I may so say, the environment of the young men who are so soon to take our places, in order that their development may be larger, higher and in better balance. Two co-operative methods have been suggested—reversing the order of study, and organizing the practical school.

Whatever the course of improvement may be, it becomes us to leave some heritage of unity to the coming race. How shall we more fitly crown a century of engineering—a century in which our noble profession has risen from comparative potentiality to living energy? And as its force is multiplied by the general advance of science, it becomes the momentum which evermore shall actuate the enginery of civilization.

DISCUSSION ON PRESIDENT HOLLEY'S ADDRESS.

PROF. T. EGLESTON.—The subject of the President's address is one of very great importance to the profession. There is, as is evident to all, a want of concord between the schoolmen and the manufacturers, but the want is all on the part of the manufacturers. The President has shown us where the knot is, but has not cut it himself nor helped us to cut it. The unwillingness of some manufacturers to allow the student or professional man facilities, not for seeing the secrets of the trade, but the every-day manipulation, amounts sometimes to almost an obstruction to education. It is certain that this unwillingness arises in most cases, either from misapprehension of the object, or from want of tact on the part of the student. It may arise, also, from a well-grounded fear that the information asked for is to be used for direct competition, in which case it is well founded, and an act of self-protection. There are those, however, who are always willing to afford opportunity and facility for acquiring knowledge, to the student, and who do not acknowledge, either in word or deed, that any antagonism exists. To their kindness and consideration, many of us are greatly indebted for substantial aid in solving some of the difficult problems. Some generous-minded men carry the desire to aid further than is wise. We, of the schools, send students to the works to acquire knowledge which will be of use to them in solving problems given them to call both their knowledge and judgment into play, not to acquire *information* which is acquired to-day and forgotten to-morrow. Some of our friends furnish sehedules, complete estimates, and lists of prices, in the most friendly manner, thinking to do a service. The result is, that the ease with which the information is acquired, leads the indolent, indifferent man to copy them verbatim, without any attempt at the very exercise of judgment, which is our principal object. The incipient engineer must decline to use such information, as his conditions are not the same; but the temptation to consider such a division of labor as legitimate, is, nevertheless, very strong. What was intended as a kindness is often no help to the investigating student, and a weight which helps to drag down the man who is willing to copy; such co-operation is not what is needed, and is no real kindness. When the School of Mines in New York was first started, it was stated, that for students who wished it, application would be made for permission to work in mining and metallurgical establishments as the highest reward of engineering scholarship which the faculty could offer; but we were soon obliged to withdraw the promise, since the manufacturers would not grant the permission. Yet it is easily to be seen that out of such men, working for a given time without wages, the manufacturer might make, if he would, men of a high order of engineering ability. The difficulty is to make the manufacturer see that it is for his interest to have men trained to think and then to see. The day I graduated from college I left the machine shop in which I had been working, to take my diploma, and went back again to the shop, and after nearly a year's work went to a technical school. For the first year and a half that I was in the profession, though in charge of very large mines, not a single important question, involving mining, arose, but there were constant questions of pumps, engines, etc., and I was led at first to think little of my technical education until it became evident that the power to use the knowledge gained in the machine shop was the result of the technical training. It is plain to any one who gives the matter a moment's thought, that no amount of learning, which can be acquired in a technical school, can give the eye-education acquired by long practice in the works, and which makes every good workman an artist. A thorough education will, however, enable one to learn the nice distinctions, which are the basis of a correct judgment, in a comparatively short time. It seems to me very doubtful whether the ability to do the manual work is desirable in the engineer. The schoolman must have experience, just as the ordinary workman, but he acquires it in a much shorter time. Success in this, as in other things, is more or less the record of failures. As a general thing, the man who has never failed has never tried. The failures of the ordinary workman, looking over long periods of years before they become skilled, are not noticed or forgotten, but the failures of the schoolmen are more prominent, because so much, that is not fairly reasonable, is expected of them. It is evident that the young engineer must supplement his education at the school by training at the works, and if his failures in properly directed efforts are compared with those resulting from the efforts of men who have not studied abstract science, it is certain that they will be found the least expensive. What the President has laid out as a practical course means that men would not get into their profession until they were thirty years of age, which, in a certain point of view, is an advantage, but could not be carried into effect in this country, where men must begin to earn their own livings about the time they are twenty-one. What is needed is to increase the requirements and advance the age of admission to our technical schools, and then to overcome the antagonism of the manufacturers, and induce them to allow the graduates to continue their study with them until they could learn enough of the professional work they must do, to feel and hear, almost by instinct, as the President has so forcibly said, what they were to do. Such a system would soon train a class of engineers not to be had now. It is easy for the schools to give men the theory, but where are they to get the practice, if the manufacturer, miner, and metallurgist is to stand by them and say, yes, you ought to swim, you must swim, but we cannot by any means allow you to go into the water until you have learned how to swim.

Mr. Joseph D. Weeks.—I cannot agree with the last speaker, that there is always this want of concord, this disinclination on the part of manufacturers to allow scientific investigations to be made at their works, or if such a want or disinclination does exist, that it is entirely on the part of the manufacturers, or without reason. So far as my experience has gone, I have found the iron manufacturers of my own city, Pittsburgh, very ready to grant facilities for investigation, and work in perfect harmony with me in making experiments. At the works of Messrs. Rogers & Burchfield I spent many days experimenting on the use of natural. gas in the puddling furnace, being at liberty to make any change in furnaces or their working I chose. At Messrs. Spang, Chalfant & Co., so far as I am aware, there has been every facility granted. But a week ago two gentlemen connected with the State Geological Survey were at these works making some very thorough experiments, and I know that a young gentleman, a graduate of the Professor's own school (the School of Mines), has been at these works for weeks.

Prof. R. W. Raymond thought the question of furnishing information to visiting students was chiefly one of courtesy and of present policy, while the question of giving a practical knowledge to students, or those who were to become such, was a much deeper one, bearing upon all the future. He thought that in many cases the questions asked of managers by students who called upon them were of a kind to embarrass one party without really benefiting the other. With regard to the suggestion of the President, that there are many considerations of practice which cannot be formulated, Mr. Raymond thought a large part of this imagined inability would disappear when greater culture was introduced among practitioners. A mind trained to think and to express thought would deal promptly with many of the customs and rules of practice, discarding one-half as superstitions or survivals from necessities now obsolete—like the rudimentary organs of Darwinism—and classifying, explaining, and formulating the rest. The speaker did not believe much in rules or practices which could not be formulated. The trouble was the inability or disinclination of practitioners to state the facts. The President's recommendations would abate this evil. With regard to the combination of technical instruction with practice, perhaps the *Steigerschulen*, of Germany, are a good example. They have furnished some excellent mining engineers and metallurgists. He thought the introduction of a preliminary course of practical training before the scholastic training would be a great benefit to the schools themselves. It would weed out the hopelessly stupid, or lazy, or otherwise unfitted, and bring to the schools a class of young men who already knew what they wanted and really wanted it. The great difficulty with the schools to-day was to recommend themselves to the public—to parents. This could be done by the original investigations and publications of the professors, and still more by the practical achievements and business success of the graduates. Anything which tended to increase the chances of graduates to obtain opportunity of advancement, and to acquit themselves creditably therein, would benefit the schools. On the part of parents, the trouble is that they are in too great a hurry, and it is imperative that we should prove to them that a more thorough and prolonged training *pays*. Too many educators are at present engaged in the ridiculous attempt to adopt from the German system of university education everything *except the time it takes*. We keep on taking out a little more classics, and pushing in a little more science in our curricula of education, leaving the whole thing to be crammed in the same space as before. The speaker thought the only solution was to give more time, and most heartily agreed with the President that lack of general culture—of the knack of study, the power of statement, the sense of order and logic—was one of the greatest defects and drawbacks to the engineer. No illiterate man (using the term in a broader sense than that of inability to read and write) could become a distinguished engineer; or, at least, the most able engineer, if deficient in the culture which gives men power over their fellows, would be the first to recognize the lack, and lament the cause that had so crippled his usefulness and stopped him almost at the summit of that fame which could be completely climbed only by complete men. One of the members of the Institute, a gentleman of high reputation as a manager of mines and works, had recently said to the speaker that he had concluded, after much observation, that a regular collegiate education was the best preparation for a technical one, and that the time so spent was well spent, and showed its benefit surely at the other end of the student's career.

MR. E. C. PECHIN.—All are agreed as to the desirability of closing the gap between theory and practice, and the future will, undoubtedly, develop the plan by which this can be accomplished; but the important present consideration is, if we cannot close the gap, may we not make it narrower. The plan suggested by President Holley is an admirable one all will admit, but its development will necessarily be slow because so radical. There ought, however, to be no real difficulty in the way of graduates of the schools becoming thoroughly versed in all matters of practice. There is an apparent obstacle, in the reluctance of the graduate to accept the prospect of a year or two of real hard work. I have had a number of applications from educated and competent young men, for positions, but in a majority of cases they wanted a good salary to begin with, and the nature of their duties must be rather genteel than otherwise. What they should say is, "I have had a thorough technical education, but I want a position for a year or two, where I can become thoroughly familiarized with practice. I only ask such compensation as will reasonably support me, and to labor for a given time in the various departments of your business."

There are very few masters who would not recognize the merit of such an application, and gladly assist so laudable a desire.

Several years ago, I accepted the services of a young gentleman, who had just finished a thorough course in the German schools. First acquainting himself with the nature of our materials, he then practically became a workingman. Whenever and wherever his services were required, he offered them, day or night, in the stock-house, or at the front. He carefully noted difficulties as they presented themselves in practice, and how they were met. The result was, that in one to two years he had mastered his business in both its aspects, and he is now, though still quite a young man, in my opinion, one of the foremost furnacemen in this country. Now, this would be my plan for making scientific practicians.

If our graduates would only cultivate that very un-American characteristic, *patience*—learn to labor and wait—their success would come sooner, and be more assured. The man who learns the art of managing men, and to this adds a knowledge of chemistry and metallurgy, may become a giant, but with all possible learning, and wanting the ability to handle labor, which can only come by the knowledge of what is required of labor, he cannot hope for any large success.

This plan embraces another important point. Laboring men, as a class, are suspicious and unsympathetic, so far as their employer is concerned. By having associated with them, one becomes acquainted with their peculiarities, appreciates the difficulties they encounter, learns how to gain their good will, and this once gained, is enduring; and when the time comes for him to control, his power is well-nigh absolute, for he has learned when to watch, or relax, when to exact or indulge, when to be liberal or when unsparing. His knowledge obtained from books and in the schools, tell him what *ought* to be done; his practical experience shows him *how* it can be done, efficiently and economically.

PROF. R. H. THURSTON.—I listened to the opening address of our President last evening with very great interest, and should have liked to have taken an opportunity to indorse the general accuracy and practical promise of his views, and his plan of securing a more complete and thorough technical education and training last night, but that I presumed that an hour would be set apart, for the consideration of the more novel and more important of the propositions, which were then made. I have been asked to open the discussion this morning, and have gladly promised to do so.

I know of no subject which has more direct bearing upon the prosperity of the profession than this, and there is certainly not one which has a greater interest for me, personally.

As a mechanic and a practitioner, I recognize the correctness of much that has been said of the "schoolmen." As one who has been accused—but I am afraid unjustly—of being one of those schoolmen, I agree in what was said of the so-called "practical" men. Undoubtedly, both classes can do more to make our methods of technical education and subsequent training satisfactorily effective. Each must, however, help the other, and the best work will be done when they work earnestly, persistently, and patiently, hand in hand, with a common and well-understood aim, and by well-settled plans, each helping the other by every means that thought, observation, and experience can suggest.

The questions how, when, and where the requisite combination of theoretical and empirical knowledge of the principles of professional practice, and the experience needed to make that knowledge most readily and fully available, are to be obtained by the aspirant, have been present in the minds of all thinking members of the profession, from the time of that great engineer, the Marquis of Worcester, who, two hundred years ago, proposed the establishment of technical schools.

It has appeared to me that we may pursue either of three distinct courses, in the attempt to effect this generally admitted essential combination. I say generally admitted, because there are members of acknowledged high standing in the profession, who assert that, however desirable in itself the education of the schools may be, no man can afford to give the necessary time to its acquisition. I am told by a friend, for whose views I have great respect—a gentleman who is justly distinguished as an engineer—that no man can attain full success in business, who has not gone into the business at so early a period in his life, that it is physically impossible to preface entrance upon his life-work by more than the rudiments of a common-school education. He insisted that only the man who began his actual practice of the profession, at an age at which his powers of observation and his ability to acquire by memorizing and by tactual experience are most fully available, can succeed.

While acknowledging that these views have much truth underlying them, I presume that very few would agree to the final deduction made from them. We may find a great difference of opinion, as to how much time the young man can spare to the pursuit of scientific professional knowledge, but we shall rarely find an advocate of its complete abnegation. The question to be discussed here is certainly not whether the young engineer shall have a technical training, but whether he shall secure it in one way or another of several proposed methods. We are not asked whether he shall have such an education and training, but how shall we give it, and when should he seek it, and where.

I have said that I would specify three courses, either of which may, perhaps, accomplish the desired result. These are:

First. That method which is most usually adopted, in which the student is given his education, and is then sent into business.

Secondly. That which gives the boy a common-school education, then sends him into the office, or the field, or the workshop, to acquire a certain amount of practical experience, business knowledge, and general development, and finally places him in the technical school, to obtain the professional education and scientific basis for a sound reputation, which can there be best and most readily given him.

Thirdly. The course which, although usually most difficult to pursue, is, if I may judge from observation and experience with a considerable number of instances, the most perfectly and economically successful. That is, a mixed course of study and practice, extending throughout the early life of the man up to his final and complete immersion in the practice of his profession.

It is possible that I may be influenced by that prejudice, which most men have in favor of a course which has answered its purpose more or less fully in their own cases, or in cases which have appeared to them illustrations of great or of even moderate success; but I believe that the boy who, with natural predisposition toward a certain branch of engineering, spends his weekly holidays and his vacations playing about the workshop, growing up in contact with the workmen, and witnessing continually all those operations which, as he becomes old enough, he learns to conduct himself, imbibing, with that wonderful accuracy and rapidity for which boys are remarkable, all the traditions and recognized principles of shop practice, learning the construction and use of tools, and now and then acquiring the art of manipulating a machine or handling a tool, I believe that this boy will most easily and perfectly secure the technics of his profession.*

The course adopted generally in this and in all countries is the first of those specified. The boy is sent to school, and is given the usual common-school education. Upon concluding this course of study, he is sent to the college, or the technical school, and a four years' course of higher education having been completed, he is sent into business at the age of twenty or thereabouts.

He has then been engaged in the work of the student all his life; habits of study have been formed, and usually he has become, to a certain extent, unfitted for the vastly different kind of occupation which is now to be taken up. He has acquired habits of study, a good memory, and the ability to utilize it thoroughly, and has learned to make logically correct deductions from properly grouped facts. He lacks usually, however, the power of quickly perceiving and promptly acting upon such perception. He probably lacks decision, has lost some of that strength of character which may have been his by inheritance, and he has none of that experience which

* See note on page 40.

is essential as character and knowledge to success in business. He may possess a great store of learning, both general and professional, a well-trained mind, a sound judgment, and all that scholastic habits and training can give him, but he lacks the no less essential knowledge of men and of things which he can only obtain by a personal contact. He cannot manage his employés without either making unreasonable demands upon them or yielding to them more than is just. He knows nothing of methods of conducting business, and cannot have become accustomed to the hard rubs which so seriously disturb the tyro, and which so often discourage him at the outset.

Habits acquired in youth are always difficult to modify in later years. His habits are those of the student, and he must inevitably find it a seriously difficult matter to acquire the peculiar and distinctive habits of the business man. Once succeeding, however, he will rarely fail of full success.

The exceptional course in this country, and I presume in Europe, is the second of those outlined. The boy goes to the shop, or the office, immediately after completing his grammar or high school course, and learns the trade which leads most directly toward the profession which he proposes to enter; or, under the tuition of some practitioner, he acquires a knowledge of the ordinary routine of work and some idea of the character of the greater problem which he may expect to be confronted with in later years.

Arriving at the age of twenty, he sees the advantage of the possession of a knowledge of the science of his profession, and he leaves his practice for a time and devotes himself to study in some technical school.

His difficulty now is to acquire habits of study and the student's power of making his own that knowledge which he finds in books, and of grasping experimental data and of collating essential facts and grouping them systematically, and of deducing from them general laws of precise definition, and of well-determined range of application. He has to reacquire the mathematician's power of basing upon a statement of accurately defined conditions generalizations which find practical application in every department of human knowledge. He has to regain that fondness for research and study of which his business life has done so much to deprive him.

On the other hand, he has learned by experience to prize knowledge, both for its own sake and for what it will enable him to accomplish. He has learned in what direction he is to expect most aid from literary attainments and scientific knowledge. He can, to a certain extent, distinguish between those branches of study which give only a mental gymnastic training and those which enable him to accomplish two objects simultaneously,—to acquire a store of valuable knowledge and, at the same time, to profit by a no less useful mental fulness of stature; becoming more of a sage and more of a man at the same time.

On the whole, I suspect that the advantages of this method more than counterbalance the disadvantages, and I have no doubt that, could this course be generally adopted, it would be seen to have an importance in the acceleration of professional progress which we probably hardly realize to-day.

Two great obstacles intervene to prevent the general adoption of this plan. The first is that conservatism which always retards the introduction of anything new, and which usually makes it necessary to agitate for at least a generation before a really great change can be brought about. Even when all are agreed on the question of the propriety of a step such as this, it is usually a long time before the public inertia is fully overcome. This is well illustrated by the fact that the necessity of technical education itself—proposed two centuries ago, and fairly inaugurated a century ago by Vaucanson, the father of the great *Conservatoire des Arts et Metiers* at Paris—is only just now beginning to be universally acknowledged by even those who are not hampered by a traditional proclivity in favor of the old Greek non-utilitarian idea of a purely gymnastic system of education.

The second great obstacle is the natural, and almost universally observed, reluctance of the young man, who has once become fairly inducted into business, and who sees opportunities opening to him in the immediate future, to give up all and to return to the school to secure advantages which his reason tells him are still more important, but which he, nevertheless, cannot fully realize, looking upon them as he does from a standpoint which does not permit him to see them as distinctly as he may in after life, when experience has confirmed the previous judgment. An active, energetic, and ambitious young man can rarely bring himself to the point of going back to the school after having once tasted the pleasures of success in business. Where this has been done, however, it has been almost invariably the fact, if I may judge from my own observation of quite a number of cases, the result is a most encouraging one. Could this plan be generally adopted, it would not only be decidedly better for the young man himself, but it would prove vastly better for the schools.

The greatest difficulty met with in carrying out a satisfactory course of technical instruction in the schools, is that of finding students who have sufficient ripeness of intellect and of judgment, and sufficient physical strength to comprehend readily, grasp fully, and retain perfectly, the principles which are presented to them. Boys are sent to technical schools without well-developed habits of study, with insufficient and superficial preparation, with minds unripe and with bodies still taxing their systems by the drain of that vital power needed in carrying on the operations of physical development. Were the last-considered plan adopted, they would come to this work, which demands all the powers of maturity, with body and mind fully developed, and with an understanding of the extent, difficulty, and importance of the work to be done which would insure vastly better performance, and the accomplishment of vastly more in the time assigned to the course. The work of the instructor would be rendered more easy and more satisfactory to both himself and his pupil. The time would be far better utilized, and the greatest good would be accomplished in the given time and by the expenditure of the given amount of time and funds.

It is in this direction, I am pleased to find, that our President is looking for higher efficiency in technical training. If the plan which he proposes, of making our larger manufacturing establishments advanced technical schools can be carried out, it will prove, I am sure, a long step in the right direction. The final portion of the work of education would be done at a time when the student has attained sufficient maturity to appreciate it, and in the midst of such influences as will most effectually impress its value upon him. I sincerely hope that a way may be found of initiating this method of tuition, and that we may soon learn just what we are to expect from it. Difficulties will undoubtedly arise, but with the exercise of care in the choice, from among the many of the few who are adapted by nature and inclination to the pursuit of the profession, and with tact on the part of the instructor and a hearty good will on the side of the manager of the works, both, hand in hand, working for the accomplishment of an object the importance of which both appreciate, there can arise no insurmountable obstacle to final and complete success. When the schoolmen and the business men of the country put their heads together in the earnest desire to accomplish such an object, their united efforts will vanquish apparent impossibilities. I think that I can see this good time coming.

Mr. E. B. Coxe.—I, to a certain extent, had a practical education before going to the technical schools of Paris and Freiberg. During a portion of my youth I lived in the Lehigh coalfield of Pennsylvania, and having a liking for such things, spent a great deal of time in the mines, machine shops, foundries, etc., of that region, and made and assisted in making surveys both above and below ground. I found this practical education of immense importance to me in my technical education, enabling me to understand the details of mining, etc., when listening to lectures by the professors and studying the subject in books. There is another point to which I would refer. The mechanics we have at present are often men who do not serve regular apprenticeships, or if they do, work only at one or two nearly perfect machines; the amount of skilled labor in the large shops is being reduced to a minimum; one man works almost all his life at one machine instead of at all sorts of jobs. A carpenter, to-day, gets almost all his work from the mills, and is more of a fitter or joiner than a carpenter, and few carpenters work more than a year or two as an apprentice. Years ago a carpenter or machinist had to learn his trade thoroughly, and hence he was more familiar with the details of his business, and would not require foremen of such ability as at present. A mechanical engineer should therefore be acquainted with details of work, so as to be able to utilize these imperfect mechanics. A machinist must now go to a country shop, where the machinery is imperfect, in order to become a thorough workman in all the branches of his trade. I feel it would have been of great assistance to me in my education if, before going into the technical schools, I had worked practically in a machine shop and foundry for a year or two.

In the paper under discussion, there is one point which seems to me to be worthy of more consideration, that is, what Darwin terms the survival of the fittest. We always hear of the one man who, from a common miner, becomes a coal operator, but of the 150,000 other miners who do not we never hear anything. One thing that cannot be learned in the school, and which an engineer should know, is how to deal with men, how to make them work in accordance with his ideas. You may make a perfect plan, and have a complete drawing, but if you cannot impress your master carpenter, machinist or blacksmith with the idea that you understand it, and that it is a good thing, it will probably not work. They will get a bolt or something else wrong somewhere.

Another trouble is the difficulty of graduates in understanding that they are not civil engineers or mining engineers, but only fitted to study the practice of those professions. The word engineer has become mixed in its meaning. I had, some years ago, a cart driver. After awhile I employed him to look after a small stationary engine. Now he is an engineer. I knew a slate-picker boy some years ago. After awhile he went with an engineer corps to carry stakes. He then became a chain man, then backsight man, and he got to use the transit instrument. Now he is an engineer. If business should grow slack, they go back to driving carts and picking slate, so that many people imagine an engineer to be a person who drives a cart, runs a transit, picks slate, or tends a stationary engine according as trade is good or not, and the popular idea of what is the proper pay of an engineer is too often based upon such conceptions of what an engineer is. Graduates should at first be satisfied if they can make their board. They will generally make mistakes enough to cover what they think their wages should be.

Mr. Oswald J. Heinrich.—I have listened with great pleasure to the remarks made by Prof. Thurston, and I am extremely glad to perceive that he fully recognizes the principles indispensable for a thoroughly practical and theoretical education of young men desirous of adopting any of the technical professions for their future careers. I rise as a strong advocate of his last proposed course, which is, according to my judgment, based on a long experience in many different positions, the only thorough plan for ultimate success. The questions involved in this controversy are of such importance, that one interested in this subject as much as I have been, for more than thirty years, may be tempted to talk beyond the limited time which can be devoted to this subject now. Permit me, however, to make a few remarks, taken mostly from my own experience during business life, which has been based upon the courses pursued in my native country.

There cannot be the least doubt, that at no time in the past has a partial education and training been sufficient for a man to fill, creditably, any position of importance in the various callings of the technical profession, and we need not expect that it will be different in the future. While partial experience alone, after perhaps a lifetime spent in a particular calling, may fit a man for a specific purpose, facts are not wanting to show that even such a man may commit great errors, or even blunders, in disregarding well-known principles which would be thoroughly understood by one of far less practical experience, but possessing a school education. It is true, a really good practical man, with indomitable energy, may succeed ultimately, but probably only after dearly-bought experience, which otherwise could have been avoided, and wasted time and money saved. This being the era when the state of cultivation of a nation is measured, to a great extent, by the most thorough use made of waste material, we may just as well say, also, that this should be extended to the imponderable items of time and brain. On the other hand, a young man of a thorough theoretical education may deliver a lecture before a set of practical men which would fill them with admiration and awe, and yet he might be puzzled by the same set of men if called on to show how to do some simple practical operation. But the result may be, that other practical men may exult over the apparent superiority of the "indomitable" spirit of the practical man (probably themselves too ignorant to judge of the cost of his experience), and sneer at the failure of the scholar, and thus bring discredit on the attempts at liberal education of mankind. A long life in various practical callings may fit a man to fill even eminent positions, and, by being cautious, he may avoid such losses as have been enumerated, while, on the contrary, the unpractical scholar, possessed, as it is frequently the case, of too much self-reliance and mere book experience, may waste time and money to overcome practical difficulties. It is therefore not surprising to see the scale often over-weighted on the practical side.

It follows naturally, particularly in this country, that preference is given to the practical man. Unfortunately, for want of thorough understanding of the subject, the choice often falls upon a so-called practical man, and educational training has fallen into disrepute. The great drawback to obtaining a thorough, practical, and theoretical education in this country, will probably be less found in the means offered, than in the unwillingness to spend the time and money necessary to obtain it.

The order of the day in this country, "to make money," and, to a great extent, judging the capacity of a man according to the amount of money he has, or is earning, will unquestionably be a great drawback yet awhile. On the other hand, it is also impossible to obtain sufficient knowledge and experience during a few years of training, and in one particular course of instruction.

Taking my own experience, I had the good luck, from my early boyhood until I had arrived beyond the years of maturity, to be alternately occupied in practical pursuits, and in receiving educational training, at schools of various grades. Until I had arrived at the age of twenty-three years I had never earned money worth speaking of, but spent my time from my fifteenth year in apprenticeships and going to various technical schools and public works. In my country, boys intending to devote themselves to technical occupations generally pursue the following plan, partially even regulated by law. After passing through the higher grades of the common school (*höhere Bürgerschule* or *realschule*) up to fifteen or sixteen years of age, where even, to some extent, Latin and Greek, but particularly modern languages, and the elements of mathematics and natural sciences form a part of the system of instruction, they are regularly apprenticed to the particular branch of business they intend to take up afterwards. As apprentices, they pass their regular time as carpenters, masons, pattern-makers, moulders, machinists at mines or furnaces, etc. Generally night schools, or schools during part of the winter (*handwerkerschulen*)—industrial schools—are visited during their time of apprenticeship, the time so spent being allowed as regular apprenticeship. They receive little or no pay during this time, according to choice and circumstances. After spending several years in this way, they enter the higher grades of the technical schools or colleges, to pass through a thorough course of scientific training, at the same time, in various ways, being constantly reminded of the practical duties necessary to be performed by them hereafter, by making excursions during the period of lectures, and during vacation visiting the public works and shops of the country. After graduating at these schools, they enter again for a time as volunteers at the different public works or private establishments, and are glad to be taken as such, without receiving any compensation, sometimes even paying for the privilege. After such a course, and proper examinations, they are only considered, even at private works, to be fit to take a subordinate position, and are often only too glad to get it.

I consider a good general education more than desirable before entering practical life for various reasons. The principal reason is, that the mind of the boy is more susceptible to mental training and exercise. During his apprenticeship, or attention to practical work, he will find out the great help he may derive from educational training. This is kept up by attending the night or industrial schools during that time. These preliminary studies, connected with practical exercise, will balance mind and body, both essential for a young man in those years of life. He will be by far a more attentive scholar at the higher grade schools, at least so far as my experience has gone, and will profit more by attending such schools than generally is the case with those who have first passed through the entire collegiate or classical course of studies. I consider this a very natural consequence of the necessary course of studies in industrial schools, they being better designed to prepare for subsequent training than the old faculty studies of law, medicine, theology, etc.

I can only say this much from my own experience, if I had not been trained and educated in such a manner as described above, it would have been impossible for me to fight misfortunes and ignorance which are necessarily encountered everywhere,—in a foreign country particularly—such as it existed in former years even in this country, and to a great extent even now.

A man who has charge of a work may be as wise as Solomon, but if he cannot show that he can lay the hand on the wheel himself, his subordinates will not confide in him. And this is particularly necessary in times of danger and difficulties, which may arise at any work. Association during his early days with the different classes of workmen, will make him more familiar with their notions, customs, and even their tricks, besides giving him at least some mechanical ability. Even if his constitution is not sufficiently strong to do a full day's work of a laborer, he may be fully able to say when a man has performed a full and good day's labor.

It is easy enough for men to manage a business where they are surrounded by good, able, and skilful men, or can get them at a moment's notice, as is generally the case in the old country, but it is absolutely necessary, in a partially settled country like this, for a man in such a position, to be in every respect capable of not only ordering a thing to be done, but also, if need be, to be able to roll up his shirt-sleeves, and show how it should be done. Men educated in such a manner may even aid the working classes, who cannot devote their time to hard study, by giving them useful instruction during their daily vacations as well as during evening lectures, and I have had opportunity to test this matter to my great satisfaction, finding our mechanics here willing enough to receive instruction when opportunity is offered.

I will, in conclusion, only say to our young men, study hard, and as long as you can, but be also not afraid, if you can improve yourself by it, to soil your fingers or your coat, even if it is at a more moderate compensation than that of a first class superintendent.

Note by Professor Thurston.—The course of instruction in which workshop and laboratory practice is given simultaneously with tuition, in the several branches of study considered essential to the proper preparation of the student for professional work is becoming well recognized in this country as the most promising of good results, and a few institutions are endeavoring to give their students such a curriculum. A number of new technical schools are about to be organized, also, in which the same course will be attempted. Yale, Cornell, and Columbia, and the colleges at Bethlehem and Easton, and others, are pursuing such methods to some extent. The Rensselaer Polytechnic Institute at Troy set a good example of truly practical technical education many years ago, and the Massachusetts Institute of Technology is working in some departments on a similar plan. The Stevens Institute of Technology has been instituted, in deference to the character and wishes of the great engineer who founded and endowed it, as a School of Mechanical Engineering.

In the Department of Mechanical Engineering of that school, over which I have the honor to preside, I have endeavored to make the combination of theoretical, empirical, and practical instruction, to which I have referred as the third possible course, as complete as it could well be done within the walls of the college, and have complemented this tuition by as much instruction among the workshops and other establishments in the neighborhood of New York as opportunity can be found for. It would be premature to report definitely upon the result of this plan to-day. Several years must elapse before the real value of a method which aspires to make young men capable of going from the college into business and soon becoming efficient aids to older practitioners can be fully judged. I can only say that I originally allowed myself five years to determine whether it would be for my own interest to continue in a work which then seemed to me one of the noblest enterprises in which a member of the profession could engage, and that period having nearly expired, I am not inclined to feel less faith than I had at first in its success, and have not lost any of the enthusiasm with which I took upon myself that task. With less kindly, persistent, and cheerful support and example than I find in the presence of my colleagues and of all the officers of the institution, I may have had a different feeling;

but where all are working together in such a cause, discouragement should be impossible.

The student at the Stevens Institute of Technology enters after very much such an examination as is prescribed in all colleges of standing, and studies four years, at the end of which time he receives a diploma, which is to be taken as evidence that he has pursued with credit a course of study which is intended to be a suitable preparation for the commencement of actual work, and for beginning to learn the business of construction. He is called a mechanical engineer, which is to be interpreted as meaning that he is considered to be well fitted to *become* a good engineer. The first two years of his course are preparative, and during that time he is acquiring a good knowledge of the ordinary college courses in literature, mathematics, and the physical sciences, somewhat amplified and modified in directions which tend toward future practical applications. The next two years are devoted principally to professional studies, and to the application, in the laboratories and shops, of the scientific knowledge previously acquired. The student works in the chemical laboratory, on analyses of ores, metals, and the materials which he expects to make use of in his future life. He spends a considerable amount of time in the physical laboratory, making experiments upon whatever has a bearing upon his professional work, and learning the methods and the use of apparatus which are needed in all work of precise measurement, and he there becomes thoroughly and intimately familiar with all that the college student usually only knows by hearsay and by eyesight in the auditorium. Here his hands become expert in manipulation, and he feels and realizes all that we, who were less fortunate in our college days, barely comprehended, and never fully appreciated.

It is only here, and by these means, and with such actual tactual practice, that the student can be taught habits of accuracy, and an understanding of the difficulty which attends the acquisition of valuable experimental data. His mental training here is by no means the least important part of the work. In mathematics he pursues the higher branches, and devotes much time to applications throughout these two years. His instruction in the drawing-room extends throughout the full course. The earlier portion of this time makes him familiar with the use of instruments, and he gives much time to the graphics of descriptive geometry, and the studies to which that forms the introductory. During the final portion of his course in that department, if he has any inventive power or ability to design, he makes original designs of machinery and apparatus. The machines for testing metals, and for testing lubricants made in the mechanical laboratory, were designed more or less entirely by students.

In the department of engineering he devotes two years to the study of mechanical science, and to obtaining a knowledge of his materials of construction. He studies the construction of typical machines, and the form, as well as the theory, of the prime movers. He goes into the workshop two days in the week, and there learns the construction, the use, and the manipulation of machine tools. He acquires a little of the mechanic's "knack;" if he has a natural sleight, he gets a good deal of it. He there sees work going on in the hands of more experienced mechanics, and gradually acquires that kind of familiarity with such matters which is so essential to his future success. He goes into the little foundry and sees the moulder at work, and soon knows the technic of that business. More or less work in pattern-making, and the other subsidiary trades is usually going on, and he picks up enough in the shop, and from the lectures on the trades, to be able to advance rapidly when he finally goes out into real work.

The mechanical laboratory, which has been founded during the past year, is doing a considerable amount of work for private individuals, railroads, and other large corporations, and has done much experimental work for the neighboring municipal, State, and United States governments. This has been of immense value in giving the students an opportunity to witness, and frequently to take part in, experimental determinations of important facts and data.

They use daily the apparatus of the engineer in the lecture-room, and become familiar with the forms of the machines for determining the tensile, torsional, and transverse strength of materials, the steam-engine, indicator, etc., and occasionally have an opportunity to take part in the experimental work which is every day in progress. Students have taken part in tests of all kinds of materials of construction, of steam-engines and boilers, pumps, and other machinery, and have used the dynamometer, the pyrometer, and all the instruments used by the mechanical engineer in his professional work. They have taken part in some very important investigations and in some valuable original scientific researches. The exceptionally complete outfit of this mechanical laboratory, and the wide range of work obtained by it in doing its work, as a matter of business for all kinds of business men, gives opportunities which, confident as I was of its ultimate success, did not, when I first attempted its organization, even imagine. It gives all these advantages, and, what is hardly less important, it pays its own expenses.

Viewing this experiment as a whole, from our present position, I can say that I see no reason to doubt that it is destined to prove itself as complete a success as its most enthusiastic friends have even hoped.

I have given this somewhat full account of the way in which this problem of technical education is taken up at the Stevens Institute of Technology, partly as an illustration of an actual and earnest attempt to carry out the method which I have referred to in my remarks, and partly for the purpose of giving those members who asked for more information a concise answer to their questions. I hope that I may have succeeded in interesting still more those who have exhibited a desire to see our methods of technical education perfected, and a high "modulus" of efficiency attained.

THE JOINT DISCUSSION

OF THE AMERICAN SOCIETY OF CIVIL ENGINEERS AND THE AMERICAN INSTITUTE OF MINING ENGINEERS,

AT THE HALL OF THE FRANKLIN INSTITUTE, PHILADELPHIA, JUNE 19TH, 1876.

MR. A. L. HOLLEY, *Member of the American Society of Civil Engineers and of the American Institute of Mining Engineers,* called the meeting to order, and said:

GENTLEMEN: The members of the Joint Committee of the Civil Engineers and the Mining Engineers, to whom was given in charge the arrangement of this series of meetings, had proposed to nominate its Chairman as President of this meeting, Mr. Henry R. Worthington, who is, as you are aware, a thorough scholar in the principles of the profession, as well as a successful practician of wide reputation in various branches of engineering. As Mr. Worthington has been to-day most unfortunately prostrated by illness, I will nominate, confident of your approval, a member of the committee who is thoroughly conversant with its deliberations, and peculiarly competent to preside at this meeting, Dr. R. W. Raymond, of New York.

DR. R. W. RAYMOND, *Member of the American Institute of Mining Engineers,* upon unanimous election to the chair, said:

GENTLEMEN: The formal title of President, which Mr. Holley has conferred upon the position of Chairman of this meeting, has struck me with a little dismay, for it was my idea that this would be an informal conference rather than an organized, deliberative, and legislative assembly, but since there is to be a machinery, and a President, I think we cannot do better than go a little further. I would like to return the favor to Mr. Holley by nominating him as Secretary of the meetings.

(Mr. Holley was unanimously elected Secretary.)

Now, gentlemen, a word or two may be well in introducing the subject of the evening's discussion, particularly as many who are present to-night may not be so well informed concerning the antecedents of this discussion as are those of us members of the Institute of Mining Engineers who were present at Washington on the occasion which gave rise to it, namely, the address delivered by Mr. Holley, President of the Institute of Mining Engineers, at that meeting, on the subject of Technical Education. Portions of that address, and of the debate that followed it, have been sent to many of you by the Joint Committee. I have myself been requested by the Chairman of the Joint Committee to hold myself in readiness, in common with many other gentlemen who were expected to express their views on particular branches of the question, but the unexpected duty of opening the whole subject seems to call for a more general statement, of an introductory character. I cannot hope to make it altogether comprehensive, certainly not so comprehensive as I should have prepared to do had I anticipated that I should be called to this position, and not left to make my individual contribution merely to the drift of the discussion.

The question of technical education, as a whole, seems to me to have two principal bearings, the first being the technical education of the working-man, and the second, the training which should be given to those young men before, during, or after their school course, who pass through our technical schools in the hope of making for themselves a successful career in the profession of mining, mechanical, or civil engineering. And the problem presented by the first of these aspects, if once successfully solved, will perhaps throw a great deal of light on the solution of the second. I mean to say, if we could successfully solve the question of the ordinary education, preparation, and elevation of the working-man, in our industrial trades, we should thereby gain a great advance in regard to the education of the superintendents, managers, and overseers.

It has been found that a technical education, and special preparation of the working-man, in the trades represented by our profession, are absolutely necessary to him, to take the place of that system of apprenticeship which is now passing away, and to relieve the working-man from the restrictive influence of the new principles that have succeeded apprenticeships, and guilds, and caste, and paternal government, and official direction in every form. I speak particularly of our own land, where these old artificial restrictions have been removed, and we stand in danger of another class of tyrannies operating upon the working-men. One of these is the tyranny of the worst phase of trade-unionism, occasionally known among us, and another is the division of labor. Trade-unions are by no means an unmitigated evil, but all will admit that they may be carried to such extremes, and be so misdirected by ignorance or malice, as to become an overshadowing and stifling curse upon the community.

It has been observed by all those who have had to do with laborers in masses, that trade-unions cannot become thus powerful for harm, among men who are themselves so educated, and so emancipated, that they have individual and personal ambitions. In all *this form* of trade-unionism—and I am sorry I have no better name for it; I do not mean to stigmatize the whole system—we have a despotism, the shackles of which hang like a millstone upon the limbs of the working-man. Good and bad workmen are put upon one level. Attempts to earn by extra work, to save, to study, to compete, to rise, are discouraged or forbidden. Against such a dead weight, crushing labor to a uniformity of despair or sullen indifference, the only available and effective lifting force is the intelligent ambition of the individual laborer.

The great extension in modern times of the principle of the division of labor is another source of difficulty, demanding the same remedy. It is not an easy thing to deal with a division of labor, by which the workman is becoming a specialist, and narrowed down to the doing of one thing which he can do, and do well, and which he is made to do all his life, if he is willing to do it. He becomes almost a part of some particular tool or process, unless he is possessed of a fundamental education that will help him to turn his hand to other things. This evil is not yet, perhaps, so deeply felt among us as abroad. Our American mechanics and laborers have a certain readiness, superficial perhaps, but still qualifying them to do in tolerable fashion, upon short notice, many things, and turn from one thing to another. This readiness, as Mr. Coxe remarked on a former occasion, can be got, not in a large shop, but in a small one, where there are fewer workmen, and greater variety of work to be done. In large establishments, you know when a man gets to do one thing well, he is often obliged to stay at it. He is so skilful and valuable a workman in the spot where he stands, that it is to the interest of the employer to keep him there. But education, awakening and arming ambition, stimulating thought, and developing ingenuity, affords an antidote to this evil. Technical education of workmen may become necessary, then, to remove the restrictive influences which I have named, and to give to them the *Open Sesame* so that they will have the possibility, if they have the ambition, to rise.

The effect of such training of workmen must be studied chiefly, I am sorry to say, in foreign countries, our own being in this respect lamentably deficient. I cannot pause, in the brief period allotted to me, to discuss this portion of the subject; but I beg to refer those Americans who take an interest in it, to a suggestive summary of foreign experience, contained in a little book on *Technical Education*, by Charles B. Stetson, published in 1874 by Osgood & Co., of Boston. That account is far from complete. It does not make mention, for example, of the *Steigerschulen* of Germany and Austria, for the instruction of working miners and metallurgists, which are among the oldest and most successful efforts in this direction. But, not to dwell upon these details, I wish to point out that such primary culture of the working-men renders possible the system of promotion from their ranks, without which they are driven to the worst style of trade-unionism in self-defence, but *with* which they are inspired at once by a laudable, honorable, and practicable ambition. These effects can be observed in the operations of the Cooper Union in New York—an institution which I need not pause to explain, still less to praise—almost alone in this country as a successful apparatus for the technical instruction of the laboring class. If the masses of our artisans, miners, and mechanics were possessed of such preparation in mathematics, drawing, and language as the Cooper Union schools bestow on their 1500 pupils, then there would be less difficulty in selecting from the bench, the lathe, the engine, the stope, and the furnace, men capable of rising to the intelligent control of great establishments.

But, it may be said, how will this help the case of those who, after years of study in colleges and scientific schools, find it impossible to gain entrance into practice? If even now, when nine workmen out of ten are mere drudges at a specialty, and incapable of promotion, our school graduates find it hard to get places, what *will* they do when workmen are educated? I reply, that they will be forced, probably, to begin as workmen, but that such an apprenticeship will be far less difficult than now, because it will be served among comrades who have learned the meaning and value of science, and who will gladly co-operate in the exchange of ideas and experience. This is not a mere Utopian dream. The fellowship of knowledge is a reality, and comes to pass whenever the vulgar scorn which the ignorant feel for learning, and the vulgar scorn which the indolent feel for labor, have been swept away by the simple expedient of making the worker study and the student work. You might as well tell me that if I were hungry, but had plenty of water, and should meet a thirsty man with plenty of bread, we could not come to an understanding and an equitable exchange, as to tell me that theoretical and practical knowledge cannot complement each other under such circumstances.

But the present practical difficulty still remains, how to give to students, either before, during, or after their course of technical study, a knowledge of practice, such as will qualify them for positions at least of subordinate responsibility, and put them in the line of promotion. In the admirable address of President Holley at the Washington meeting of the Institute of Mining Engineers, a plan was suggested which merits not merely discussion but trial. It proposes a period of practical instruction in the shop or works, after the student has received a good degree of preliminary culture, but before he has begun his strictly technical studies. I shall say a few words on this plan as applicable to metallurgical industry, for it will not have escaped the attention of those who heard or who have since read Mr. Holley's address, that its suggestions more particularly concerned the mechanical engineer; and the question at once arises whether his plan is necessary, feasible, and advantageous to the metallurgist.

1. In the first place, it must be recognized that metallurgical processes involve more and more the arts of civil and mechanical engineering. They are carried on upon a larger scale than ever before, and hence they require heavy constructions and complicated machinery. For a great part, perhaps the greatest part, of the duties of the metallurgical manager, whatever is necessary to the training of a mechanical engineer is necessary to him.

2. It is more easily practicable for the student to gain a general acquaintance with many mechanical tools and processes in metallurgical works than in the machine shops themselves. For the repairs and alterations of machinery around such works give a variety of illustrations which the original construction in large establishments, where labor is rigidly divided, does not offer to the student or apprentice in any one department. But, on the other hand, the actual elementary knowledge is perhaps more difficult to obtain, because the operations at metallurgical works require, but do not train, skilled mechanics.

3. As to the manual part of construction—the masonry of furnaces, for instance—it can be readily learned at the works, where constructions are almost always in progress in some form or other. But I think that if the student has had beforehand sufficient training to understand the use of working drawings, and to be able to make such drawings, he will profit better by this practical experience.

4. As to the purely metallurgical portion of the work, the charging and operations of mills and furnaces, the sorting and testing of ores, the manipulation of products, etc., there is no doubt that a practical familiarity with the facts as they exist will facilitate the comprehension of theories and general rules. But the student cannot go far, and ought not to go far, without the aid of chemistry; and I am inclined to think that a general knowledge of that science, and some experience as an assayer, will prove most useful preliminaries.

5. Is it feasible to introduce young men into metallurgical works on the footing of favored employés, that is, employés who are not to be kept indefinitely at a single branch of the work, but who are to be passed from one department to another, until they shall have acquired familiarity with all, as a preparation for thorough study of the sciences underlying them? I think it is feasible in some kinds of works, and scarcely so in others. In gold and silver stamp-mills, for instance, the formal arrangement contemplated by Mr. Holley is not necessary. A young man of intelligence, activity, and trustworthy character, has only to make himself useful in any capacity, as laborer, bookkeeper, or assayer, around such an establishment, to gain the *entrée* to all its mysteries. And permit me to say here, that in small mining and metallurgical enterprises generally, I think the bookkeeper stands the best chance of promotion to the manager's place, when it becomes vacant, if he is ambitious, and makes good use of the opportunities to gain a knowledge of the whole business, which no other position offers in equal measure. This is of course not always the case. In very large establishments the bookkeeper may be closely confined by the mere clerical drudgery of his place.

In conclusion, I wish to emphasize what I remarked on a former occasion, that whether technical instruction be preceded or followed by manual practice, one thing must precede both, to insure the highest success in any profession, and that is general culture. For success is a social matter; it depends upon a man's influence over men. Knowledge of facts and laws in nature will not achieve it.

The most thorough metallurgist or engineer needs to be able to make other men recognize his ability. Nay, long before he can acquire thoroughness, he is dependent upon other men for every chance of practice. A liberal education gives power over men; and the technical education which gives power over matter will be twice as easily gained, and twice as effective when gained, if it is grounded upon the mental discipline and the moral strength of a culture wider than its own.

The more one observes the careers of men about him, and the more one wrestles with difficulties of one's own, the more profound becomes the conviction that a young man makes a great mistake, who, because he is going to take a technical education in engineering, deliberately decides that he will not have any general culture to begin on. I am not speaking of the men who, struggling against cruel necessity, make their way honorably and effectually, in spite of early disadvantages. Such often win a place among the greatest names. But the reason is very simple. It is just the same reason as makes the Indian a hardy son of the forest. Excessive exposure, hardship, insufficient food and clothing, do not make men hardy; they merely kill off the men who are not hardy, and those who survive must be the vigorous ones. Poverty, ignorance, isolation, difficulty, are not elements of strength; they are obstacles over which strength, and strength only, can triumph. Infinitely better they are than the luxury that drowns ambition and breeds swamp-gases of indolence and vice; but in themselves they are hindrances. A man who is truly a man will not be enervated, but enlarged and stimulated by liberal culture.

I would appeal to no one sooner than to our self-made men for a hearty recognition of the value of such preparation. They have felt the lack of it too keenly not to wish for their children a better chance. Now, with due caution against the waste of time, I cannot doubt that a general culture, though it may not be the quickest preparation, will lead to the best results. I remember the remark of a man of great success and quick observation, who assured me that if his son would become a metallurgical engineer, he would put him through college first, and let him begin his special studies afterwards. I am not prepared to say that an entire college course is necessary, or that it is the best preliminary course, though I have a high opinion of it, but something equivalent to it, or to a part of it, that is, what our German cousins give to their young men in the *Gymnasium*. They give to them a liberal culture in the beginning;

and it is a very remarkable thing, that amongst the many skilful metallurgists and mining engineers from Germany with whom I have had the pleasure of becoming acquainted, I have found a large proportion who had learned Greek and Latin, could perhaps even play on some musical instrument, and were widely acquainted with literature.

Finally, we must recognize the fact that individual character is, after all, the decisive element in success. We may devise plans without end to facilitate the manufacture of skilful engineers, but the men who have fidelity, honor, virtue, courage, and that genius which has been well defined as the power of application, will make their way surely to the top, either by the help of our arrangements, or in spite of them all; and of these born and bred leaders of the profession, those who have the broadest culture, other things being equal, will stand easily first.

I will now call upon Mr. Thomas C. Clarke, of the American Society of Civil Engineers.

MR. THOMAS C. CLARKE, *Member of the British Institution of Civil Engineers, member of the Society of Civil Engineers, and of the Institute of Mining Engineers.*

MR. CHAIRMAN: All will admit that education by books and education by practice are both necessary to a perfect system. The practical question that we have to discuss is, "What or how much shall be taught from books, and what or how much by practice?" Let us once satisfactorily answer this, and the rest is easy. The time to be occupied in each and the order of study is of minor importance.

Civil engineering is what is generally termed one of the practical sciences. A practical science is the application of scientific laws in various departments to serve practical ends which govern their selection and application.

Thus civil engineering being the art of scientific construction (of construction in accordance with the laws of nature, and not in ignorance of or against them), selects from the abstract sciences mathematics, and applies its branches of geometry and algebra to surveying and measuring. It selects from physics, chemistry, geology, mineralogy, etc., whatever is necessary for its purpose, and cares for nothing else. The pursuit of abstract science meets with its respect, but does not command its attention.

The study of the abstract sciences should be thoroughly pursued before we begin to practice, for that is the only time that we can properly give to it. The application of these abstract sciences should be left until we come into practice, because men engaged in practice are best able to judge properly how that application should be made. Practice is constantly changing, and the applications of science have to be modified daily.

Moreover, the power of observation will not be thoroughly acquired until the student goes into active practice, while in the schools he deals with the symbols of things, and often does not recognize the thing itself when he sees it.

The schools should confine themselves to teaching principles only. That moment they attempt to teach the application of these principles to practice, and endeavor to teach their students how to design cars, engines, and other machines, permanent way, bridges, roofs, dams, and such pieces of construction, they either are ten to twenty years behind the age, or else they incur the ridicule of practical men by teaching how to make something of no use.

I have already expressed my views before the American Society of Civil Engineers as to the order of study that I would recommend. This, however, is of minor importance as compared with the previous question of *what* to study.

I would recommend that the engineering pupil get as sound a general education as possible, including the principles of the sciences. Let his early education be rather that of general culture, developing his mind, strengthening his powers of observation and judgment, teaching him to generalize. This course he should, if possible, pursue up to the age of eighteen or twenty. Before that age the mind and body are not generally sufficiently developed to endure the physical hardships of engineering. Then let him spend several years in practice in the machine shops, in the field, in the drafting-room, and in the office. Let him learn to deal with men and things, and to understand the conduct of affairs. Whether he will return to his books again depends upon what sort of a man he is.

I believe that all men, or nearly all men, from the natural constitution of their minds, fall into one of two classes.

They are either the men of executive ability, the practical men, *par excellence,* those who have a natural talent for affairs, the organization of labor, and the direction of men; or else they are the men of science, the investigators, the men who are hungry for knowledge, and will learn the reason why. Very rarely one man unites both qualifications. James Watt did, and so did Professor Morse, but such men are rare.

If the young engineer belongs to the executive class, having once plunged into practice, he will probably never go back to his books. But the other kind of man will do so, either by himself or in the schools. When he will have found out exactly what his deficiencies are, and he will be able to judge (much better in most cases than his professor can) what it is desirable for him to learn, you may be sure of one thing, he will study the principles of science, and pay very little attention to their application as taught in the schools. He will not spend his time over the pages of Rankine, learning how English permanent way was made twenty years ago, before Mr. Bessemer was heard of. Whatever he studies will be of value to him, and no one can judge of that better than he can.

One thought more and I have done. To all classes of engineering students let me point out the immense value of acquiring and fully understanding the *scientific method.*

This is, first, the art (for it does not come by nature) of observing facts and acquiring data; second, of observing the relations of phenomena and of drawing conclusions therefrom; third, of verifying those conclusions by observation and experiment.

Robert Stephenson, in alluding once to the vast progress of modern engineering, in which he himself had borne so distinguished a part, said: "We found it a craft, and we have left it a profession." That is to say, it had been put on a scientific basis, and by the use of the *scientific method.* This, after having been applied to the construction of railways, is now beginning to be applied to their management, and the results are remarkable, and promise to be more so.

To give an instance. The invention of the recording dynamometer, by Mr. Dudley, enables us to determine the resistance of trains at different rates of speed, and consequently of the number of foot pounds of energy required to move them, which is only another name for cost.

It was formerly the opinion of practical men that *ten* miles per hour was the most economical speed for freight trains. The instrumental observation shows that eighteen miles per hour is really the most economical speed. Perhaps ten miles was right in the days of iron rails, but now with steel rails and better rolling stock eighteen miles is the proper rate.

But how long would it have taken the man who was not possessed of the scientific method to find this out? And consider the value of the information thus obtained. It means postponing the cost of constructing additional freight tracks, until the freight business has reached a point nearly double that which it was supposed would require them.

Other instances might be given of the value of the scientific method, but this is enough.

CAPTAIN DOUGLAS GALTON, F.R.S., *of London.*

MR. CHAIRMAN AND GENTLEMEN : I have so little knowledge of the American system of education for engineers, that I fear I may add very little to your information on this subject, except by perhaps giving a short account of the arrangements which have been made by our Indian government for the purpose of endeavoring to create a department of civil engineers for India. The Indian government has recently, that is to say, within the last three years, undertaken the construction and supervision of a much larger share of the public works than it had previously done. It decided to make new railways under its own officers; therefore it required the assistance of a large body of engineers. The first idea was to obtain the assistance from the civil engineers of England, and to induce young men of that class to go out and accept service under the Indian government as civil engineers; but the civil engineer is, as you are well aware, a very expensive creature, if he is trained sufficiently in practical skill to be of real value.

The government of India were not disposed to pay for the article the amount which was required to purchase it. After the failure of that experiment, in consequence of the inadequacy of the terms offered, the Indian government determined to create a school of their own, and they have established, at a place called Cooper's Hill, near London, a college of civil engineering. The pupils for this are selected by competitive examination, and they are then subjected to a course of technical education. This course is admirable as a theoretical course; the young men are taught, as far as possible, a certain amount of each branch of engineering. This is the theory, but the real difficulty lies in giving a practical education to young men in the schools or in colleges. You may teach them by models the art of bridge building, or the principles involved in the construction of revetement walls; or you may cause various sorts of work, as a plaything, to be executed by or under the eye of the student; but it is perfectly impossible that those difficulties of engineering which are of daily occurrence in actual works, and which are the things that give the engineer his real practical knowledge and education, I say it is impossible that these can be taught in any schools. The young men, when they leave Cooper's Hill, proceed to India, and are placed under government engineers. It is almost too soon to say what the result of this will be, but I think it is not difficult to predict that it cannot have so favorable a result for the government as the result would have been if they would have been willing to pay for students who had been educated under our leading civil engineers, the price which would have attracted them, and it was not a very large price that would have attracted them.

There are two causes which militate strongly against the success of such a system.

First, the career of these young men depends upon their being regarded with a friendly eye by their superiors; that is to say, that promotion more or less depends on favoritism.

Secondly, government service is antagonistic to inventive genius; its leading principle is that the younger men should conform to the precedents established by their superiors. No encouragement is given to the development of new ideas.

Of all professions the civil engineer's is the one of which new ideas are the life's blood; it is, therefore, the one which should be least trammelled by being made a government department.

It is noteworthy that the colonial office, which has jurisdiction over certain of our colonies, and is independent of the Indian office, has recently generally selected the civil engineers for service in the colonies from the civil engineers of England, and this plan has proved very successful at no inordinate cost.

In some of these colonies where considerable works go on, as railways and roads, they have an organized service of civil engineers, which has been selected from pupils of the civil engineers in England, and their chief advisers in the selection have been our leading civil engineers. Moreover, the system at Cooper's Hill is very similar to that which had long prevailed in India, namely, that of employing the officers of the Indian engineers on public works. These officers used to have the same sort of theoretical education as the Cooper's Hill students—perhaps not quite such a high one—they had not the opportunity of getting that practical knowledge which is necessary to an engineer, and they therefore had to acquire the practical knowledge at the expense of the government, and many serious errors were made in construction from that cause.

It is a far more expensive thing to pay for the education of engineers by means of their failures, than it is to purchase the educated, practical engineer, at the highest price in the market. I was not aware that the question to-night was to embrace so large a field as the President shadowed forth, or I should have taken more pains to give you further information on another branch of technical education, namely, that of the technical education of working-men, because in England we have a recent example of the efforts made to improve the technical education of the artisan, by that eminent scientific machinist, Sir Joseph Woodworth, who has put in trust a sum of two hundred thousand pounds for scholarships, to be given to the working-men who will distinguish themselves in a technical examination, and these scholarships are for the purpose of allowing a certain sum yearly to students for a limited number of years, to enable them to have an opportunity to improve themselves in their education.

I will briefly lay before you a few points as to the sort of education civil engineers receive in England. First they receive a liberal education at some school or college, and then the rule is that they go as regular or apprenticed pupils to a leading civil engineer for a specified time, and they pass from that state of pupilage into assistants, and finally they settle themselves on their own basis. Mechanical engineers invariably serve as pupils in the shops of some of our leading machinists, and they, as a rule, pass through all the mechanical processes in these shops in detail, as well as through the drawing office, and the model-room, as a part of the course.

Mr. Robert Stephenson once said to me that he considered that he owed a great part of his success in life to his father, old George Stephenson, having insisted upon his working in the model-room of the foundry for a whole year, or two years, before he would allow him to proceed to other classes of work. He said that gave him such a power of judging of the forms of things, that it was of immense benefit to him. I would only add, in conclusion, that however much education you may give a young man, you can at most only teach him how to learn. The real education is what he gives himself after he leaves his instructors. It will always depend upon the young man himself whether he makes an engineer or not.

MR. COLEMAN SELLERS, *Member of the American Society of Civil Engineers and President of the Franklin Institute.*

MR. PRESIDENT AND GENTLEMEN: Mr. Holley some time ago, inclosing a copy of a circular which I hold in my hand, asked if I would speak on the subject of the Technical Education of Engineers, as such education relates to machine shops and foundries. In preparing mentally what I should say this evening, I naturally thought only in relation to the question here asked, "Shall a course of instruction in works precede, accompany, or follow that of the technical school?" but so much has been said to-night about the technical education of workmen, that I can hardly avoid saying a few words on that subject.

More than fifty years ago, a few earnest mechanics of the city of Philadelphia met together, and feeling that they needed some means of instruction not obtainable in the ordinary course of their lives, founded what they called the Franklin Institute of the State of Pennsylvania, and in time came to occupy the building in which we are now holding this meeting. Almost the first act of their existence was the foundation of a school for teaching drawing. For very many years there were no schools in Philadelphia to teach mechanical drawing, other than the one existing in this building.

From the Franklin Institute drawing school sprang the Women's School of Design, which has since grown in usefulness, but if at the present time you examine into the free school system of Philadelphia, and I imagine of some other cities, too, you will find that it is with the utmost difficulty that the teaching of drawing has been introduced. It has been declared by the authorities in some instances that there shall be drawing taught, but practicably there is none, even with such a rule existing. Now, Mr. President, I hold that the very foundation of all engineering practice is the knowledge of that language which is the universal language of the world, the language of the pencil, a language expressing in form what we desire to explain.

If this language of the pencil could be taught to all children, even if they did not get quite so much grammar and quite so much of some other branches in the schools, they would be better fitted to be workmen, and would make better workmen, too, but, unfortunately, everybody has not the same idea. The tendency of the school system of the city of Philadelphia tends directly towards making young men accountants or traders, or, it may be, giving them an idea that they are to be doctors, or lawyers; and not one word is taught them in the direction of the industrial arts. They do not have a chance to learn the meaning of the commonest terms of the machine shop or the foundry, or of any of the trades. Now the want of this system in the public schools underlies the whole trouble of technical education in America. We do not in the common schools impress children with the desirability of earning their living by handicraft, nor do we train their minds to fit them to direct the hands of others in the industrial arts. Least of all do we give them any hint of what they want if they are to be engineers. Some technical education is needed to prepare them for a higher practical capacity on leaving the schools. The high schools now, it is true, teach many things that are very useful for them to learn, and too much cannot be said in praise of what is really the People's College of Philadelphia.

Now, in regard to the question of what shall be done with our sons, assuming that we would like them to be engineers, to follow in our footsteps. I am sure there is no father who listens to me now, who has not had this subject very seriously brought to his attention. For my part, I speak feelingly and from my heart, when I say that it is a question which cannot be always satisfactorily answered. I considered it gravely for a long time before I concluded what was the best thing to do with my own sons, and I may say what I have done with them is what I think is right for others to do with theirs under similar circumstances. I cannot but indorse Mr. Clarke's advice to secure for our boys in their education, as broad a foundation to stand upon as possible. I am sure this cannot be done by sending them to a public school only; they should have some college education; colleges properly organized will grow into favor as training schools for engineers. I am not sure that the ordinary university course of Latin and Greek is the best, but even this has its advantages, provided the young man can spare time enough before entering upon his life work to obtain some scientific training besides. I really think it would be a good thing for our young men to go through a thorough collegiate course, and then take something of a scientific course. But the end seems to be more nearly met by establishing in all our universities, scientific schools. Where such schools are founded, the student will enter college, and may work for two years with those who are studying a regular classical course. They will learn much that will be of great use to them in after life; they will master language so as to obtain the power of intelligent expression of their thoughts; they will understand the use of words better than those of us who have not had these early school advantages. The ability to tell other people precisely what they think, and what they want to impress upon them, is in itself a power. Their minds will be expanded, and in the course of two years, they probably may have found out what they want to do in the future, though I do not think that any boy, and when I speak of a boy, I speak of one under the age of eighteen or nineteen, can possibly tell what will be the proper place for him to occupy in this world.

There is a place, I presume, for everybody, and each will after awhile drift into his proper place, and may become distinguished in that place if he can fill it properly. In illustration of this drifting into place, I would add, that if you will go upstairs, and look over the roll of members of the Franklin Institute, through its fifty years of existence, you will find names of members there, with their occupations opposite to their names as they wrote them at the time they signed. You will find one name there that I well remember, and opposite that name you will read the word "bricklayer." He is now a very celebrated architect. And so with the name of another member of the Franklin Institute, who is really a very able mechanical engineer, and quite a good constructive engineer. Had he signed his name to that constitution, when he was some years younger, you would have found opposite that name the word "shoemaker." Now, that man at that time, I have no doubt, was an excellent shoemaker, but he had in him the making of an excellent constructive mechanician. Since then he has found his proper place. If we could determine exactly what our boys are likely to be, what would be the best for them to do, then we could look over the course beforehand; but as we cannot do that, I think the position held by Mr. Clarke is exactly the one that should be pursued, and that is, to give them as broad an education as possible, but in giving them this education, I would like to have one idea pretty well emphasized, and that is, that a college will not teach them anything at all of practice, but that it will teach them what they want very much to know, and that is, how to study. They will surely learn methods, the methods of thought; they will by application study lessons, obtain a correct knowledge of the habits of thought, and methods of scientific investigation. Thus they will learn that they must never try two experiments at the same time, if they want to arrive at the bottom of any matter. They will realize the advantage of making one experiment at a time, and, eliminating all the errors, finally come down to the hard facts which they are seeking for. This is one thing they learn to do. Now it makes no difference if they cannot see how they are going to apply this when they come into the workshop. My word for it, they will see its value when they are once at work, and engaged in digging their own way through sometimes very rough experience.

Then it is they will turn back to the knowledge stored in their minds at college, and readily apply it to their work. But to reap the full advantage of all this, they must come out of college feeling that they are technically, perfectly ignorant; they must be willing to learn of practical men what constitutes the practice in machine making. If a young man can have a proper education in that way, and be induced afterwards to enter the shops and work his way up, beginning at the very bottom, and never shirking any work because it is dirty or disagreeable, you may depend upon it, the young man will grow up to make his mark in the world. These qualities obtained by education may be developed into administrative ability, and show their usefulness in making the boy a man capable of leading men. The trouble with most young men who have gone through their technical course is, that they feel they have spent a great deal of time in acquiring knowledge that should make them full-fledged mechanical engineers at once. They enter the shops, and feel sadly disappointed at meeting the young men who have no education, and are vastly their superiors in mechanical skill. This position, however, holds with them only for a very few months, because their habits of thought make them consider the reason why this or that thing is done, and when they have special training they rapidly go beyond those who have not had the same advantages. It is safe to say, that a young man, after passing through college properly, and having a good sound education, who determines to succeed in the workshops at any hazard, will in two years make himself so valuable in the position that he occupies, as to be elevated by his employer into something higher.

Now I say that I thought of this very deeply in the case of my own sons, and I did precisely, and I am doing precisely what I have just told you. I was not at all surprised, when I found my eldest son, after leaving the university, accepting a position in the workshop a little better than a common laborer. He commenced by chipping the scale out of the boiler. I tell you it was the best thing for him, because he made a beginning at the bottom, and did not shirk his work; it was as much as to say that he was willing to learn all that could be taught him in the shop, and he rapidly rose to a position higher than many who had been longer at work, but who had less book learning to back them.

It is impossible to make engineers out of pupils who have not engineering ability. There must be something in them that will compel them to take it up as a profession, and succeed in it. I am now clearly of the opinion that as it is not in the power of most young men to take the college course, and then afterward to take the technical course; that it is far better for them to obtain what scientific knowledge they can in a good college, or in a technical college where something else is taught besides the exact sciences, where they can be taught the languages, not the dead languages but the modern languages, and taught at the same time rhetoric, composition, and all that will enable them to express themselves; and by all means let them have a good sound basis of mathematics before they venture their education in the workshops. Then when they have entered the workshops there will be time to acquire technical education without schools. I have no doubt that many who have been liberally educated, have, after entering the shop, felt the want of some technical education, and have broken away from the shop, and gone into schools to learn. They felt the need of obtaining more knowledge, and that the time they spent in the college or school was not sufficient.

I do not think it advisable as a rule, however, to take the boy from the workbench and send him to school a second time. I have in some instances noted the effect of such a course upon young men to be disadvantageous. If the boy has left school too soon, and feels afterwards the want of more knowledge, it is well enough, if he can, to return to his studies, but such return makes sometimes a disadvantageous break in his habits. I look upon it rather as a means of mending a defect in education rather than a course to be pursued as prearranged with an object. By not attempting to teach too much "practice" in the schools, time is left to give a good grounding in generalities, which cannot fail to be of use in any walk of life, and which can be better acquired when one is young. The practicing engineer has not only to master his profession, but he must learn how to place himself and his works before men so as to be seen of them and appreciated by them. He requires a very extended knowledge; all learning will at one time or another be of use to him; and habits of study, which will enable him to continue a student to the end of his days, will the more readily fit him to rise in his profession, and make him a leader among men.

THE CHAIRMAN.—What Mr. Sellers has so well said with regard to the ability of trained men, possessed of liberal education for mastering new problems or new branches in applied science, is very admirably illustrated in the case of a gentleman upon whom I shall now call, a fellow-laborer of mine, and one of the old members of the Institute of Mining Engineers, who was once a very good lawyer, but who made up his mind, when necessity called upon him, that what anybody else could learn, a lawyer should be able to learn, and who took off his coat and went to work at his own blast-furnace, and became an excellent blast-furnace manager.

I call upon Mr. E. C. Pechin, of the Dunbar Iron Works.

MR. E. C. PECHIN, *Member of the Institute of Mining Engineers.*

MR. CHAIRMAN AND GENTLEMEN: In being called upon to speak at this point in the proceedings, I feel very much like a Lazarus sitting at Dives' gate.

After the admirable addresses of the chairman and of the gentlemen who followed him, so little remains to be said that I feel as if my part must consist in simply picking up the crumbs that have fallen from richer tables.

I can only discuss the question from one standpoint, that of a blast-furnace manager, as I am not sufficiently familiar with other branches to give an opinion of any particular value.

I cannot too heartily indorse the remarks of the gentleman who has immediately preceded me as to the vital importance of a liberal education as the groundwork of a professional career. *Modesty* is one of the most charming and valuable of human characteristics. As a young man lays broad and deep the groundwork of knowledge, as he enlarges the range of his acquirements, as he gathers information from varied sources, as his mind becomes widened, broadened, and liberalized, so does he feel his own insignificance, and how little he knows compared with the illimitable and exhaustless fields of the unknown lying before him.

The most unbearable of creatures is the youth who, thinking he knows everything, practically knows nothing; and if, by a liberal education, we can help to secure *modesty*, certainly a vast deal has been done, not only for the individual, but for those with whom he is to be brought in contact. With a liberal education as the groundwork, a subsequent technical education becomes an element of great power.

As far as regards blast-furnace management, the schools can effectively *precede* actual practice. A young man who looks forward to the position of manager of blast-furnaces of the present day cannot have too thorough a technical education, as his duties will be of the most varied character. Heretofore very many establishments have been content to make iron out of purchased materials, but the tendency now is to concentrate the mining of ores and coal and the making of coke and iron all under one general management, greatly increasing the duties and responsibilities of the manager, and demanding of him a larger knowledge.

Though he may have efficient subordinates in various departments, he must be able to understand and check their work, and he will find it extremely useful to be something of a mason, bricklayer, and carpenter, largely a surveyor, engineer, and mechanic, and well up in the general principles of chemistry and metallurgy.

Hence it follows that the course in the technical schools cannot be too varied, and the best plan for an earnest student to pursue, while thoroughly mastering the particular branch selected as a specialty, is to get the groundwork of as many collateral branches as possible, as he will find in his after-life that nothing will pay him a larger dividend than varied knowledge.

After the schools, *practice*—not the practice that many of our scientific fledglings expect—a good salary and comfortable quarters, strictly business hours and social enjoyments—but earnest, hard, often disagreeable, and, it may be, poorly paid work, commencing at the bottom and going resolutely through, until all the details of his business are familiar to him. All the knowledge of chemistry and formulæ in the world won't help a furnaceman if his scales are out of order, if the furnace has slipped, or water is going into his crucible. The most perfect furnace and the best stock can come to grief if the materials are not properly filled at the tunnel-head.

I would start the graduate in the office, to become familiar with the method of keeping accounts. Many a hardworking iron-man has been insensibly ruined by faulty bookkeeping. Thence to the weighing department, to find out how to receive supplies and ship products. Defective scales or false weighing may knock the bottom out of an otherwise fair balance-sheet. Thence into the labor gang of the furnace, helping, filling, firing, keeping, until he is acquainted with every-day furnace life, the petty mishaps, and the more serious complications. From his fellow-workmen he can learn a vast deal that will be immensely useful in his after career. He will discover how many work intelligently, how many work honestly but ignorantly, and how many stupidly or wilfully shirk. At any rate he can find out the peculiarities of the average working-man, how distrustful and unreasonable at one time, and how generous and confiding at another. He ought to learn how to win the working-man to his future support, and if he does this he will possess the power to snatch many a victory from the jaws of apparent defeat.

To sum up, first college, then the technical schools, then thorough practice, to make a good iron-master, provided always that the subject has the right stuff to start with. As Mr. Sellers's remarks have suggested, "You can't make a silk purse out of a sow's ear." If a youngster has shown himself a good student, practice will soon prove whether he will make a good manager. If the necessary qualifications are wanting, he will soon find his true level, but if they are present he will, by adopting the course suggested, rise surely and rapidly.

Quite apart from the financial and commercial results which will inevitably follow, his life will prove a success to *himself*. The habits of the student will cling to him through the busiest practical life; his desire for knowledge will never slacken; the scientific research of the world will become his property; and he can appreciate and practically apply what others have done and are doing; he can understandingly avoid the blunders, and intelligently appropriate the discoveries of his fellow-men, and possibly and probably will in turn benefit others by the results arising from the combination of a constantly ripening intellect and an ever enlarging practical experience.

Thus we shall have a thoroughly skilled mechanic, and at the same time, what is quite as important in my estimation, a broad-minded, intelligent gentleman, equally at home in his overalls among his men, or in his dress-suit in the society of the cultivated, refined, and intellectual.

COL. W. MILNOR ROBERTS, *Vice-President of the American Society of Civil Engineers.*

MR. CHAIRMAN: "The Joint Committee of the American Society of Civil Engineers and the American Institute of Mining Engineers on Technical Education" having requested me to join in the discussion, I cheerfully accede to the request, the discussion to be chiefly upon the following questions:

"I. Should a course of instruction in works precede, accompany, or follow that in the technical school?

"II. Is it practicable to organize practical schools, under the direction and discipline of experts, in engineering works?"

Being less familiar than many of my brother engineers with the details of modern technical teaching, my best excuse for entering at all upon such a discussion may be found in my acceptance of the proposition, that "in a multitude of counsellors there is wisdom."

I look back over my own experience, beginning many years ago, in the practice of engineering, when there was no such thing in this country as a technical school in any branch of engineering—in fact, before engineering could be considered established as a profession. There were engineering works here long before that, and, of course, engineers, but the time had not yet arrived when engineers were in sufficient numbers, or in sufficient demand, to support regular schools of technology. On this very site where we are now assembled, and in the old Carpenters' Hall, back of Chestnut Street, about fifty years ago, I was taught in a night school, architectural drawing (under the auspices of this same Franklin Institute) by the late John Haviland, an eminent architect of that day. This, combined with mathematical training (in the daytime) in the school of Joseph Roberts, then the best mathematician of Philadelphia, aided by his interesting weekly experiments in chemistry and philosophy, was the germ of the technical education of the time. It was prior to the era of railroads; hydraulic works, canals, etc., constituted the chief engineering employment of the few engineers then in the United States.

If I am not in error, this very society, the Franklin Institute (which has so kindly tendered us the use of its halls and library, and the most liberal personal attentions to both of our societies), was the first substantial practical movement in the direction of technical education, though beginning in a comparatively crude manner. Yet I am well satisfied that the lectures that I heard in the rooms of the Franklin Institute, which I always attended when I could, laid a good foundation, which no length of time could impair. The respected professors and teachers of that day have most of them long since finished their labors in this world, and may be laboring, for aught that we know, in another and more refined sphere of action. They have gone, but their works live after them. In weakness and inexperience they laid the foundation walls of the Franklin Institute, a little more than fifty years ago, and now nearly all its early members have departed, but the Institute has thriven and grown to man's estate, and it would be very difficult to estimate its real value as a factor in the technical education of the American engineering mind. Certainly it has been very great, partly in its preliminary training of youth, partly in its valuable practical lectures to young and old, partly through its numerous important scientific committees, and largely through its monthly publications, which permanently embody a large amount of technical knowledge, valuable not only to the student, but to professional men of every grade of experience. I have, on the spur of the moment, interjected these reminiscences, in the hope that they will not be deemed entirely irrelevant to the subject in hand.

In regard to the training of young men of the present day for the profession of engineering, there are some preliminary considerations which it appears to me have not yet received quite as much attention as they seem to deserve, and they are becoming more important every day. It is but a few years since engineering in general took rank as a profession, and we look back to a yet shorter period when schools of any kind undertook to teach engineering as a specialty in any of its present branches, military, civil, mechanical, and mining; but modern progress has been so rapid, and operations demanding skilful engineering experience have so increased in number and magnitude, that already these several branches have become in themselves sciences, requiring for their proper development and treatment appropriate, distinct schools. All of these branches of engineering have assumed definite shape during the present century, and chiefly during the last fifty years. Not that there was no military, civil, mechanical, and mining engineering prior to this century, but that the business and wants of mankind had not yet advanced sufficiently to necessitate the formal organization of such professions. Since the introduction of the railroad and telegraph, with steam as the great motor for all kinds of machinery and transportation, an almost unnatural impetus has been given to all branches of engineering, and now we have schools and colleges specially devoted to the teaching of our youth, not merely in the preliminary or elemental parts, but in the more advanced and practical portions of these sciences, and annually there are now numbers of young men, who have passed through the prescribed course in their respective institutions with sufficient credit to entitle them to their diplomas, sent forth to take their chances for employment in some one or several of the branches of engineering.

Under such a comparatively indiscriminate system, two results follow, one, the spoiling of intellects which might be much more advantageously employed in some other business of life; the other, the general public loss from incompetent engineering service. It may be said that the professions of divinity, law, and medicine are exposed to similar mistakes and similar consequent unfortunate results. This is probably true, and perhaps the same general remedy may be applicable to all. It is certainly true that all the students in any profession cannot arrive at the head. Some must be lower in the class than others, and some must be near or at the foot. Shakspeare says, "some men are born to greatness, and some achieve it;" it may be added now, "by diligent labor and untiring perseverance." Hence a course that may be appropriate and advisable for one student, may not be so for another.

Aptitude for a particular profession, and even for a particular branch of it, if it be engineering, has not yet been adequately studied or cared for by parents and guardians of children. The idiosyncrasy of every being is his own, and it is unchangeable, and it is possible, with proper care and judgment on the part of the parent or guardian, to learn its nature in their children or wards, before determining the particular plan of their education. There are undoubtedly legal, medical, theological, architectural, mechanical, and engineering minds, which, if properly trained in their natural channels, may become not only useful, but distinguished, and which, if directed and persistently held in the wrong channel, may prove to be abortions.

Regard being had to the aptitude of the individual, the particular course which may be advisable in his special case, should depend measurably, and perhaps considerably, upon the branch of engineering which seems to be best suited to his idiosyncrasy, whether military, civil, mechanical, mining, etc. While there may be a few instances of men whose minds and pursuits enable them to shine and to be more or less successful engineers in most of these branches, as a rule, those who attempt to figure in many will be likely to fail in some, if not in all. At the same time, an engineer in any department should of course be more or less familiar with the leading principles of all the cognate branches. Yet, with every year's augmenting experience in the world, the necessity for these separate branches will become more apparent, and a larger share of attention and time will be demanded in the individual pursuit of each. Science grows with the growth of the world.

It may be necessary, or at least advisable, when considering the subject of the proper method of training the young engineer, to have special reference to the particular branch of engineering he intends to follow. For any branch there must of course be a proper foundation, to the extent of a good English education (if German and French are added, it would be decidedly advantageous), and a ready use of figures, and of mathematical principles, to precede both technical engineering study, and practice—this in any branch.

In mining engineering particularly, the student to be reasonably accomplished, should also understand chemistry, as well as geology and mineralogy. In the other branches of engineering, chemistry may not be so necessary or important, although it is a kind of knowledge which is useful to all engineers. An accomplished civil engineer should be familiar with mechanical engineering, and not ignorant of mining engineering, though he need not, necessarily, be an expert therein; it could hardly be expected of him. His chief or highest duties are not embraced in either of those branches, and his principal requisite is ready, sound judgment, and the more this is strengthened and confirmed by experience, the better for his employers as well as for himself. Sound judgment can never be wholly the result of education, either technically in the schools, or in engineering practice, because it does not always accompany knowledge or even experience. For civil engineering, the teaching and training in those higher schools, where this department, with the use of instruments, is a regular course, the student can learn all that it is necessary for him to know, before taking a very subordinate position in a regular engineer corps in the field, where he would still have much, very much, to learn, which cannot be conveyed to him thoroughly in any other than this final school. One who would be a first-class mechanical engineer, as soon as he has the proper foundation of schooling already mentioned, should get at once among mechanics and mechanical practice, joining to his practice an intelligent study of principles, and gathering to himself the technicalities as he advances. Something, too, depends upon the special branch of mechanical engineering he may wish to follow. Mining engineering during the present century, and especially during the last forty or fifty years, has become a highly important branch of modern engineering, and only in technical *schools of mines*, or among mines, can it be thoroughly learned. The proper foundation referred to above is, of course, essential to the satisfactory advance of the student. In addition to this a taste for the profession, as well as the talent to enable him properly to exercise it, are indispensable to distinguished success.

I am less familiar with modern mining engineering as a science, than with any of the branches of engineering, and do not therefore feel competent to advise mining engineers as such, but from the drift of my remarks it will be observed that in a general way my experience would seem to favor first, the laying of a good foundation—a sound preliminary education, and secondly, going to work upon it as soon as possible thereafter, in the particular branch of engineering the student intends to pursue. If that preliminary education could have the great advantage of such schooling as the President of the Mining Engineers, Mr. Holley, has indicated among *actual works*, it would surely be the very perfection of a foundation.

After all, however, those who would be engineers, or their parents or guardians for them, should have some reliable assurance that some kind of engineering, as a profession, is likely to be the best for the particular individual, and then, the kind of engineering should be determined upon and constantly kept in view during the preliminary or preparatory education.

I can speak freely in regard to civil engineering, from long experience, that in the United States at least, it has become of a very mixed nature, and that of those who have latterly entered into it, many have been sadly disappointed and very few quite satisfied. Many who without due consideration have rushed into the profession, would have done better for themselves in almost any other business or profession. Civil engineering has been largely overdone. Perhaps the same general remark may be more or less applicable to the other professions, although circumstances peculiarly favored a large accession to the ranks of civil engineers.

Since, however, the profession of engineering in its several branches may now be regarded as established, and as special professorships and special training in those branches respectively, in schools, must materially conduce to the permanent benefit of the pupils, and consequently of the profession, it would appear particularly in mining engineering, that a course of instruction in works should *accompany* that in the technical school. In mechanical engineering very much the same system would seem to be highly desirable.

In military engineering, in the national school at West Point, this is precisely what is done. Practice, so far as may be, goes hand in hand with technical teaching. In civil engineering it is practicable, only to a certain limited extent, when compared with the wider field which must at a later period be occupied.

It may not be equally practicable to organize "practical schools under the direction and discipline of experts in engineering works," in all of the branches of engineering, but in mining and metallurgical engineering it seems to me to be quite practicable and desirable, likewise in mechanical engineering. In civil engineering the real school is largely in the field, beginning with the rapid preliminary explorations of lines of canals, or railroads, or projections of water-works, etc., extending through the processes of provisional and final locations, up to the planning and construction of the various works and structures appropriate to the particular improvement. Those who by great experience become experts in civil engineering, are usually too closely occupied in the professional conduct of works to take an active or controlling part in the business of educating younger members, excepting as above indicated, by having them in their corps on active duty of some kind. The day may come in this country when civil engineering may assume a somewhat different shape, but at present it appears to me that the polytechnic schools in our country, in which civil engineering is a leading feature, furnish adequate training for young men desirous of becoming civil engineers. Of course the more thoroughly the teachers are themselves grounded in the practice as well as the principles of civil engineering, the better it is for the pupils, though it may be well to consider that the most expert, and the most experienced in practical engineering, are not necessarily the best teachers. There are men peculiarly well adapted to shine and succeed as teachers of young engineers, who would not be selected to take the responsible practical management in particular lines of civil engineering, while there are many instances of good practical engineers who would be likely to do no honor to a technical professorship.

MR. ASHBEL WELCH, *Member of the American Society of Civil Engineers.*

MR. CHAIRMAN: I suppose all agree that the future engineer should remain in the school or college till he is eighteen or twenty years old, and should get all the general education he can, up to that time, before he begins his professional education.

But experience shows that a *long* course of technical study, preceding and unaccompanied by professional practice, is highly inexpedient. I propose to glance at some reasons why it is so.

The object of the philosopher is to attain scientific results; the object of the engineer is to attain directly beneficial ends by using those results. One gets up the tools, the other works with them. Engineering education should therefore aim at readiness and skill in the application of science, rather than at scientific investigation or accumulation. The habit of mind good for one, is, when carried far, bad for the other. Too long study of science without applying it in practice, induces a habit of allowing knowledge to lie dormant in the mind, of regarding it as end, not as means, and to a greater or less extent, produces incapacity for applying it.

Many years ago, a foreigner was found on a work under my charge, plying the shovel and wheelbarrow, who had acquired a large amount of knowledge by years of study at a continental university. But though he knew so much, and was so expert in abstract science, he was unable to make any earthly use of it. He could not be taught to apply it to anything. In learning a giant, he was a child in everything else. This may be an extreme case, but it illustrates the tendency of all study and no practice.

On the other hand, practice keeps one on the *qui vive* to know the reasons for doing things, and the laws that operate. Old George Stephenson, for example, picked up a great amount of knowledge, because his practice made him hungry for it, and enabled him to assimilate it. Men of practice came to know, by what looks like intuition, things that science teaches other men only by a long course of reasoning. The habit of applying knowledge is more influential in inducing men to acquire it, than the possession of it is in inducing them to apply it.

A habit acquired in the practice of turning knowledge to account, is more valuable than a large amount of knowledge. In Franklin's time, myriads of men had much more knowledge than he had, but his habit of applying it made his little more valuable than their much.

Of course a man must have *some* science before he can apply it. But this he can get at school, or college, or in a *short* course at a technical school, while his mind is yet flexible. But a long course, reaching to a more mature period of life, fixes in the now rigid mind, a habit unfavorable to engineering success.

It can hardly be doubted that instruction in works should, when possible, accompany that in the technical school; just as the young lawyer or doctor learns to practice while studying.

Too much time spent on scientific abstractions and refinements (however useful such things may be to the philosopher), is more than wasted by the engineer; it unfits him for practical usefulness. Napoleon said La Place was good for nothing for business; he was always dealing with infinitesimal quantities.

A general ought to have been a captain in his younger days, but if a man continues to perform captain's duty up to the age of fifty, he is not likely to make much of a general. So an engineer should begin low down. But the student should not be kept long in acquiring mere manual skill. What he wants is mental skill. He should be practically familiar with iron, but it would do him little good to be expert in making horseshoes.

It is only early practice that can teach the self-reliance, energy, and enterprise so essential to an engineer's success.

The engineer has to do with cases where the laws of nature act in different directions. Science alone cannot often give the exact resultant of those forces, sometimes unknown, often separately incapable of measurement. Experience must give the habit of estimating what allowances should be made for unknown actions and unknown quantities. Men of science once told the engineers to make fish-bellied rails, so that they should not break in the middle. The foundry laborers that broke up pig-iron could have told them that the rails would break close by the supports. Science teaches that with perfectly elastic bodies the angle of reflection is equal to the angle of incidence; practice teaches that with material bodies as they are, it never is. Time was, when for such reasons, there was some truth in the saying, that the stability of a structure was inversely as the science of the builder.

The best engineering is that which in the long run accomplishes the purpose at the least cost. The engineer should not be a mere engineer, looking only at engineering results, for then he will lose sight of their subordination to economic results. In this way so many parties have been ruined by splendid engineering. Actual practice, where money is scarce, is the best way to impress this on the young engineer. He should learn not to do, propose, or advocate anything that will not pay.

The engineering books in our schools often contain rules deduced from the experience of other countries that are wrong here. For instance, we were erroneously taught by English practice that fish-plates for rails should be only fourteen inches long, and that they should be of iron because steel would break. A young graduate once made a road for temporary use, and made the drains of cut stone, because his French text-book said that was the way to make them. It reminded me of what a man breaking stone once said: "For making roads I'd rather have one good broad-backed turnpiker than all the engineers this side———" (Sandy Hook). Early practice can alone prevent these errors from taking too deep root.

Leaving this part of the subject, I now propose to call attention to another and very important part of an engineer's education, using that word in its widest sense. I mean to inculcate as an essential feature of an engineer's character the strictest integrity.

One-eighth of all the wealth in the United States is railroad property. It is controlled by a few hundred men. Of these many are, or were, engineers. They were selected partly because their training fitted them for railroad management, and partly because they had a reputation for integrity. It is of the highest importance to shareholders and bondholders to find men perfectly trustworthy. No sane person will confide his property to a man of whose honesty he has not some assurance. Many a share would to-day bring ten or twenty per cent. more than it does, if it was certain that the management of the property it represents was, and would continue to be, perfectly honest. It is probable that character for integrity will soon be a much more influential element than heretofore in securing good positions and good salaries. The insurance the shareholder gets by the employment of a man of known integrity is worth many times his salary.

It is therefore, to say nothing of higher reasons, of the greatest importance to an engineer, to have a reputation for honesty. It is a very essential part of his capital.

But only a few individuals can be very widely known. Most men are therefore judged, not so much by what is known of them personally, as by the character of the class to which they belong.

Hence it is of the greatest importance that engineers, as a class, should have a character for stern integrity. One of the aims of these societies, and of the schools for the education of engineers, and of the individuals interested, should be to secure this character for those whom they recognize as engineers. To have the reputation they must have the thing itself; and they must keep out, and when it becomes necessary, weed out, those that are found not to have it.

That this is practicable appears from what has been already effected. I will say nothing of our societies, though I think it is safe to assert that to be a member of either is *prima facie* evidence of integrity. I prefer a reference to the graduates of West Point, whose high tone has, with a few melancholy exceptions, carried them with unsullied reputations through temptations to dishonesty by which most other men have been overcome.

Whether or not this end can be fully attained, or whatever should be the exact steps towards it, it should always be kept distinctly in view. One uppermost thought may here be presented, not suggesting any action by these societies, for that would be outside of their province, but for the consideration of each individual member. Most of us doubtless agree with the solemnly expressed opinion of our own venerated Washington, that the only sure foundation of morality, and of course of the high-toned integrity that is so essential, is the Christian religion.

PROF. FAIRMAN ROGERS, *Member of the American Society of Civil Engineers.*

MR. CHAIRMAN: I shall naturally be disposed to treat the subject under discussion from the point of view of an instructor. Formerly Professor of Engineering in the University of Pennsylvania, and subsequently taking an active interest in that particular department of the University, the consideration of the subject has necessarily been constantly forced upon me. I have given it much attention, with the result, that apparently a great many of us have reached, that it is a very difficult question, and that it is almost impossible to come to a positive decision upon many of the points which present themselves.

We all so far seem to agree upon one point, however, and I most heartily indorse the expression that has been given to the opinion, that a broad, strong basis of cultivation is necessary to enable the average man to obtain a reasonably high position in the profession. We must leave out of consideration those men who, by wonderful genius and energy, or by extraordinary advantages of connection, attain a position which is not within the reach of the ordinary man.

In technical schools, such as the Towne Scientific School of the University of Pennsylvania, we should aim at teaching the underlying principles of science generally, and the application of special sciences to that branch of practice to which the student is directing his attention, with the greatest possible thoroughness. There is in my opinion a disposition in all our schools and colleges to teach too many subjects, and an attempt to spread the time and effort of the student over too wide a range.

That cannot be attended with satisfactory results, and I believe that we shall do best by adopting a small number of studies, very carefully taught, and reiterated in such a way, that the fundamental principles which underlie all the arts and sciences shall be thoroughly understood by the student, and shall have become an integral part of what we might call his mental life.

In my opinion the time at the disposal of the student, before he enters upon the actual practice of his profession, can be best employed in the schools, without practical work, further than the small amount which may be necessary to fix in his mind the theoretical principles which have been presented to him, provision for which can be made by very simple workshops and laboratories under the control of the professors. Beyond that, I doubt very much whether the attempt to combine practical with theoretical instruction gives an equivalent for the time spent, and I believe that the interruption of the course by a year or two years of practical work in a shop or in the field would not in the main be attended with any satisfactory result. Habits of continuous study are formed with difficulty, and should not be broken in upon until the time arrives for them to be exchanged for habits of work.

The industrious student may, with undoubted advantage, spend his vacations in each year of his study in such observations of practical matters as he may have opportunities for, a course which will result in fixing in his mind very strongly the principles presented to him by his text-books and his instructors.

There are so many things that can be taught properly only in the regular progressive methods of the schools, such as pure and applied mathematics, and mechanics, and the like, that there seems to be every reason for embracing the opportunity which can never occur again, and requiring the student to devote his time exclusively to such subjects.

Once launched into the hurry and excitement of practice, the young man finds the systematic pursuit of such knowledge difficult, if not almost impossible.

Other subjects in the same category are those based upon the digested experience of many investigators, which, though to a certain extent empirical, and wanting the logical completeness of mathematical investigations, must be adopted as embodying the principles which underlie practice.

Belonging to this class are the laws of the regimen of rivers, the action of currents, and the flow of tidal streams, or the various matters of shop or constructive practice which a man must know thoroughly at the very outset of his career, and which have been reduced to form by the labor of hundreds of individuals.

We may be assured that the young man who goes out into the world with an entirely thorough theoretical education properly given to him by competent, progressive, live instructors, will be in a position in which he cannot make serious mistakes, and from which he will surely in the long run distance those competitors who are less thoroughly prepared.

The absence of an exact knowledge of the principles which underlie practice is, I think, painfully apparent in the larger number of so-called practical men, and while we constantly hear the practical man regretting that he has not had the opportunity of obtaining that theoretical knowledge which appears to him to be so desirable, we rarely hear the man whose theory has preceded his practice complain in the opposite direction.

In a case that came under my notice some years ago, a portion of a new building was covered with a half-span, lean-to, iron roof, from which was suspended a light ceiling which hid the framing from view. With the first heavy fall of snow of the succeeding winter the roof fell in, and the removal of the ceiling disclosed a curious condition of affairs which accounted sufficiently for the accident.

The contractor being sent for, expressed unbounded surprise, and insisted that, as he had put up several *whole*-span roofs, from the same drawing, of eighty feet span, this *half*-span roof of only forty feet ought to have been unnecessarily strong, and it was difficult to explain to him that, by cutting his drawing in two, he had converted an inch and a half round iron tension rod, which was amply strong, into a compression piece which was useless. In my opinion no properly educated graduate of an engineering school, in his first year of practice, could possibly make *that* mistake, and yet I am certain that similar things, coming out of well-known workshops, will present themselves to the minds of many of my hearers.

A similar case is stated to have occurred in England, where some wise individual attempted to give additional support to a whole-span iron roof which was thought to be rather light, by inserting a row of columns under the centres of the principal rafters, with the same satisfactory result.

In individual cases the precise method of education may be modified by the peculiar connections of the student giving him extraordinary facilities in certain directions, but I would sum up my remarks by saying that the best time for a young man to acquire a systematic knowledge of the fundamental principles of his science is while he is in the school, and while he is attending to what is usually called his education, and we may feel assured that he will rapidly overcome whatever temporary disadvantages he may labor under in the outset of his career for want of practical knowledge, and in a thorough and scientific manner apply those unchanging principles which have sunk into his mind and become a part of his professional nature.

THE CHAIRMAN: Gentlemen, I think it is only fair that the individual who caused all this trouble by his words, and set the two societies by the ears, and brought all of us together here to debate and discuss, and shed various lights on the subject that he has made prominent, should have a chance along with the rest of us, in the general fray, and therefore I shall presently call upon Mr. Holley to make his contribution. I ought to announce, that after Mr. Holley's remarks we shall adjourn, to meet again in this place at 10 o'clock to-morrow morning; and before he begins, I must assume the chairman's privilege of speaking without leave, more than his share, in order to emphasize one or two things which this whole discussion has brought out. It has been a very remarkable discussion in some respects. The unanimity of feeling in one particular has been manifest, namely, as to the value of broad and general culture. This is very agreeable, because it shows that all the engineers are in favor of that thing; yet I may say that the parents in this country as a class are just the other way. When an American father talks of putting his son into any special profession he says, "I am not going to send my son to college, because he is going to be an engineer. I will take him out of college." He says, "My son is going to be a merchant; I will take him out of college;" and parents are all the time pulling their sons out of college because they are going to go into some special line. As I say, the tendency on the part of fathers is exactly contrary to the tendency on the part of experts. When a man happens to be both an expert *and* a father, like my friend, Mr. Sellers, then the boy gets a wise preliminary training. But he has put his boy into his own line, and he understands what is necessary in that line. It is not difficult for a hen to bring up her chickens; it is when the hen hatches a duck that the trouble comes in; and it is the fathers who are ministers, doctors, and lawyers, who have seen some young men rise to wealth perhaps in engineering, and have got a vague notion that it would be a good thing to make engineers out of their sons, it is such fathers who are apt to think they must take off a portion of the general culture, because they fancy it does not require so much general knowledge to enter the engineering profession. They may be the soundest men on other subjects, but they know nothing about engineering. This subject has been condensed by my friend, Dr. Wedding, in a single phrase which was carelessly thrown out in his remarks, and which, perhaps, he did not consider specially significant to us: but it was very much so. When sketching the career of a young man coming into life he said, "Well, then, at about twenty-five or twenty-six years of age, it is about time to begin to look about for some means of supporting himself, so that, when he gets to be thirty years of age, he has a place in the world." We would think it was too much to expect a young man to wait until twenty-five or twenty-six before he actually cuts loose from the parental hearth. It is to the other extreme that we go, and the young men of this country set out in life with too hasty a preparation—crammed, instead of trained at a much earlier age.

One word more in regard to the suggestion of Mr. Welch concerning the value of general practice and general acquaintance with all the branches of practice on the part of the engineer. Everybody knows how hard it is to keep up in all branches and be master of all the branches. It used to be possible years ago, but it is not possible now. No one can do anything like keeping up with the progress of the science of the present day. Humboldt was, perhaps, the last living man who kept pace with science in his race. It is not because we have not men of great mental power at the present day, but because the circle of knowledge has grown beyond the capacity of the single mind. Yet there is always a certain ability of comprehension in a man who has become a master in one department. At the same time he may keep up a general acquaintance with other departments, and be an appreciative listener and helper to the specialties in other lines. I only mention this because it is worth emphasizing that our technical societies are admirable agencies for maintaining this general acquaintance by agreeable social methods. As to the matter of integrity, spoken of by Mr. Welch, I would suggest that it ought to include more than financial integrity. Of those who are employed in a subordinate as well as in leading engineering positions, the fewest number have to handle funds without any possibility of control. Where you trust a man with a dollar once, or a hundred dollars once, you trust him one hundred times with some duty, the conscientious discharge of which involves really a larger amount. We as engineers should have that peculiar variety of integrity which consists in strict military obedience. If this is wanting at the scales and at the tunnel head, we shall not know what is going into the furnace, and all our science elsewhere will be in vain.

I now call upon Mr. Holley to close this session.

MR. A. L. HOLLEY, *Member of the American Society of Civil Engineers and the American Institute of Mining Engineers.*

MR. PRESIDENT: I shall refer only to one of the topics mentioned in the circular, and I shall endeavor to maintain the proposition that technical education in works should precede technical education in the school.

Let us first clearly appreciate the proposition, and in so doing, answer the first and hastily considered objection that is often made to it. The proposition is not to send the uncultured youth into the works and then into the school, for the uncultured youth is fit for neither the works nor the school; but it is to send the graduate of the high-school, the young man of good general culture, well founded in mathematics and in the principles of physical science, first into works, that he may there learn from nature and from the actual possibilities of labor and cost, the farther value and direction of technical studies. What we are so often told, that the mere common-school boy, or the machinists' apprentice, or the average mechanic in any department, will stick to his low place and in his narrow circle—that he would neither seek the technical school nor utilize its advantages if he did seek it, may be perfectly true, but it does not affect the system under consideration. In this system there are three distinct elements; if one is wanting the whole fails.

1st. The high-school, the liberal general culture to acquaint the mind with the history of human endeavor, and more important still, to discipline it in the methods and habits of scientific thought. Without this fundamental culture, the average man—I beg of you not to keep calling up in your minds the exceptional cases—without this culture, the average man will always keep in his low plane and his narrow circle, whatever his occupation. To him, practice will teach nothing but skill, and the strictly technical school will be as an unknown tongue. I presume this much will be universally admitted, certainly it cannot be doubted, when we see the thousands and tens of thousands of young men, who never rise above the lowest stratum of mediocrity, not for the want of average ability, but for the want of that ambition which general culture inspires, and for the want of scientific *methods*, without which high achievement is but the exceptional prerogative of genius.

This much being admitted, the second element of our system, is practice in works or in the field. At this point, two diametrically opposite courses are advocated. We hold that practice is now necessary to fit a man for the strictly technical school. Our opponents hold that the strictly technical school is necessary to fit him for a course of instruction in works. In order that there may be no misapprehension of our position in this matter, I take the broad ground, and I call upon the majority of men who have made engineering what it is in this centennial year, to bear me out in the assertion, that at *this stage* of a young man's culture, the strictly technical curriculum of books, is not only useless, but positively noxious. I do not mean mildly wasteful of time, but positively disseminative of error, the marks of which can never be effaced. Let it not be supposed that we undervalue the technical school, on the contrary, if time permitted, we would quite as strongly oppose the equally absurd practice of neglecting at the proper time, the abstract principles, the formulated results and the refined methods, which can only be acquired in the school, and which can only be acquired in a manner helpful to practice and to progress, when a touch of nature has revealed their significance.

To illustrate the noxious effects of beginning a technical course with abstract principles and with whatever is formulated in books: It will not, I think, be questioned by experts, that the grand elements of success in the management, for instance, of metallurgical manufactures, of machine building, of steam propulsion, of constructive engineering—elements far exceeding in importance the mere calculation of strains or the mere analysis of materials, both of which can be bought like the materials themselves—that the grand elements of success in making all these engineering enterprises *pay*, are an intimate knowledge; 1st, of what, and how fast, and how long, and at what cost, the tools and implements and processes at hand will *do* what they are set to do; 2d, of what iron, and steel, and brass, and materials in general may be relied on to stand, not under uniform stress as laid down in the tables, but under friction and heat and vibration and multiform stresses, and all the variations of quality due to imperfections of manufacture; 3d, of cost as modified by all the risks of waste and loss by varying refractory materials, by breakdowns, by facilities of transportation; 4th, of economy as promoted by keeping all departments of manufacture and construction in harmonious operation; 5th, and most important of all, of the management of labor, founded on the knowledge of what labor may be expected to do under all the influences of trades-unionism and bonuses and lockouts. These are things which the young man who proposes to educate himself for leadership in engineering may not indeed know at once, with that nearness of knowledge which makes the expert, but unless at the very outset of his technical course, in the novice's first period of susceptibility, his mind is moulded to grasp the great facts of his profession in the concrete, as existing in the subject itself, then the abstract as imparted in the schools will acquire a precedence in all his mental processes, from which he can never be emancipated. This is the reason why the men who actually manage engineering works to-day, are so largely sprung from the ranks, despite their want of scientific training, while so much noble culture is year after year following after the phantom of abstract principles as dissociated from the subtle and ever-changing aspects of nature.

I would further express the belief, contrary to the opinion of an expert with whom it is almost presumptuous to differ, that this concrete knowledge of the profession cannot be formulated any more than—if I must use a strong illustration—than Ovid could adequately formulate the art of love. No book can impart it, no formula can give it vitality; nature with all her subtle processes must prepare the ground, and then the schoolman may plant his seed with hope of harvest. Shall the average student satisfy the conditions of progress with the book, when Tyndall and Henry, Faraday and Agassiz, threw away the book and penetrated the inmost tabernacles of nature?

The student may indeed study in works, after he has completed his course in the school, but his mind has then been so saturated with general principles and theories, that he is positively unfitted to appreciate the great considerations of cost and policy. Let me repeat this fundamental truth—the first impressions received by a mind virgin to technical training, especially the impressions ingrained by a course of books, will inevitably mould its whole future. If a man at first imbibes, for instance, the idea that certain formulæ of strength cover all cases of engineering construction, or that the known principles of chemical affinity will give him working conditions under all changes of temperature, then he comes to regard this teaching as the law, and all variations from it fill him with doubt, perplexity, and discouragement; while if he first learns from practice, that the complete law is as yet unwritten and unknown, that the first principle of success is to lie close to nature as she reveals herself both in inanimate materials and forces, and in the developments of humanity, he will not be baffled and disappointed at apparent exceptions to general laws; he will rather be stimulated to search the books, in order to draw conclusions from wider ranges of facts, and to avail himself of the methods and conclusions of other observers.

The teaching of books, however important at the proper time, is nevertheless one-sided, partial, and incomplete; the teaching of objects and phenomena can impart no error, but it reveals all those conditions under which theory and general conclusions must at last be tested.

That men of general culture who have been trained in practice, do not only seek the school, but hunger and thirst after the wider facts and formulated results laid down in books, is a notorious fact, and the reason is that such men alone can appreciate and utilize the school.

This bring us to the third element of our system; the technical school *after* some practical training has opened the mind to its advantages, and protected it against misconceptions. This subject has already been discussed.

To sum up the whole matter, let the student first survey the ground, and then plan his structure; let him first observe that all so-called laws in physical and natural science are in constant process of modification, as nature gradually reveals herself, and then he will spare himself the mortification, and he will spare the public the damage, of trying to square nature to the book.

TUESDAY, JUNE 20TH.

THE PRESIDENT, DR. RAYMOND, called the meeting to order, and introduced Mr. Boller.

MR. A. P. BOLLER, *Member of the American Society of Civil Engineers.*

MR. CHAIRMAN: The mere fact of our meeting here on this occasion, as practicing engineers, to discuss, out of our experience and observation, the subject of "Technical Education," is sufficient evidence that technical schools fall short of accomplishing the objects for which they have been instituted. Before giving, however, my views upon the specific topics named for the discussion, I am constrained to state briefly my conceptions of the principles of education, and in a measure the conditions limiting their application.

Education, in general terms, aims to discipline the mind in correct methods of thought, through the study of the best fruits of human experience and knowledge, and so to better fit the recipient for taking part in the general struggle for existence. It has two principal divisions, viz., liberal education, or the education of culture, and special education, which undertakes to select some one branch of knowledge, and pursuing that to the exclusion of all others, to make the student valuable to that portion of society having need of such special knowledge. The former is preparatory, while the latter is the culmination of educational training. In the one, knowledge is secondary to mental discipline; in the other, mental discipline is secondary to knowledge.

The "technical school" is a special school of comparatively recent establishment, growing out of the demands of a peculiarly commercial civilization, and is valuable to the community just in proportion to the degree to which its graduates have mastered the facts and bearings of scientific observations, with their application to useful ends. Thus far those under whom technical schools have been developed have been to a great extent theorists in educational matters, and so keeping their minds fixed on the abstract side of the question, have lost sight of the realities of every-day life, and the limits imposed by a very complex state of society.

An unquestioned difficulty in blocking out a satisfactory method of technical study is the absence of any national system of progressive education, by means of which students may be brought to the same plane of preparation for entrance to the technical schools—a deficiency which will probably always exist in this country, owing to the peculiar nature of our political institutions, and which must be faced in any remodelling of methods that may be undertaken hereafter.

Theoretically, the technical school is highly practical; practically, it is highly theoretical. A strong infusion of practical knowledge, that is, the knowledge derived from observation and experiment, with a corresponding reduction of pure theory, would serve the end to be attained by a technical education much better than as now conducted. It must ever be borne in mind that the office of the technical school is not to produce scholars, professors, or mathematicians. Beyond a certain point, theory, for the mass of students, is a waste of precious time (particularly that region of theory built up without fact or experiment), the limit being defined by the requirements of actual practice. Some few, and only a few, are capable of absorbing the highest developments of mathematics or physics, and such will follow the bent of their genius, curriculum or no curriculum. It is notorious, that as a rule, the tone of mind of that class of students is such that they are illy fitted for business relations, their natural calling being in the purely intellectual field of original research; and in this place they have a very important niche to fill in the economy of divided labor.

The principle of divided labor in the operations of society is of increasing importance yearly, and no technical scheme of education can meet the wants it aspires to fill, without giving full weight to this law. There is, in these later days, too much to know in any one branch of knowledge, for one mind to compass it wholly. There can be no more Humboldts, and the day has long passed since the engineer was supposed to be equally capable in all departments of the profession. Any one department is more than a life's study for the best of us, and while it is right and proper to have a certain amount of generalized information, eminence is attainable in only one or two directions. It is as an *expert* in some one subdivision of professional knowledge that the modern engineer makes his mark, whether as a hydraulician, mechanician, a manager of mines or of railways. In the enthusiasm of scientific devotees they are too apt to base all methods and systems of instruction on the idea that the mass of students will pursue science for the sake of science. They forget that the profession of the engineer partakes more of a business character than any other, that it is anything else than an ornamental calling, and that the civilization which has developed it is one of gain and personal advancement.

There are almost insurmountable difficulties, such as attend no other branch of special education, which present themselves when we attempt to outline a complete system of technical education based on practice or experimental research. There is no way by means of which students can be graduated engineers, in the same sense, for example, that schools of medicine or of theology can graduate doctors or parsons. The former have the adjuncts of the hospital and dissecting-room, while the latter, being schools simply of opinion and dogma, require nothing beyond the teaching of men trained to transmit these special opinions and dogmas. To graduate technical students on relatively the same plane as is done by the schools of law, medicine, and theology, has thus far been the ambition of the theorists who are responsible for the present system of technical education. This is shown by giving degrees instead of certificates at graduation. It is not theory that makes the engineer, but practice, in the shop, in the field, or in the mine, and the difficulty of incorporating such practice in the curriculum of instruction appears to be insurmountable, except, possibly, in some minor divisions of engineering study, such as surveying and minor geodetical operations.

It has been suggested that the "shop" should in some way be added to the regular course of instruction, either preliminary to, or succeeding, the theoretical training, and Mr. Holley has even proposed, with some degree of enthusiasm, "to establish organized schools in the various existing works." I have very little faith in the accomplishment of such an adjunct, desirable though it be, at least in any systematic manner. There is no doubt that many students, if so recommended, will avail themselves of opportunities to enter the "shop" in an individual capacity as apprentices, but to expect that any scheme can be devised to incorporate as a system, the workshop idea in the curriculum of a technical school is, to my mind, utterly out of the question. Mr. Holley's recommendation, persuasively as he puts it, is too dependent upon mercantile philanthropy, and involves a harmony of action between the schools and the owners or managers of works hardly to be expected this side of Utopia. There are two ways, however, in which the essential benefit of shop association can be had by the schools, and they are either to make a certain preliminary artisan or field training essential to matriculation, or to require a certain amount of such experience after graduation, as essential to a degree. The former course would seem the natural inductive method, as the student would commence by observing facts and phenomena, not only of material objects, but regarding the methods and habits of men with whom he will have so much to do. To a mind filled with the facts of nature, theories founded thereon have a value of application and a depth of meaning impossible to one who is purely book-prepared. To observe first and generalize afterward is the philosophy of modern progress and discovery, and it would seem to be simply following that inductive system in giving an affirmative answer to the first part of the first proposition of Mr. Holley's circular. There is, however, a serious objection to such preliminary practical training in the mechanical branches of engineering, and that is the moral effect upon youths at probably their most impressionable age, associated as they must more or less be with the rough working-man and mechanic. The injury done to their character might outweigh any benefit they derived from the opportunity of practical observation. If this objection is regarded of too serious a character to be overlooked, the only alternative is a post-graduate practical course, for I throw out of consideration the toy workshops of the schools themselves. Mr. Holley is quite right when he says that a school of engineering practice, based upon faculty control, "can be nothing less than a vast and successful establishment for construction and operation in nearly all the departments of engineering." That, of course, is impossible.

Many young men do substantially take a post-graduate course, which might be made almost compulsory on all students, and thus indirectly part of the school system, if degrees were withheld until the student gave *proof* that his post-graduate experience was such as to entitle him to be ranked as an engineer. The present method of graduating young men as full-fledged engineers is not only a perfect farce, but it also gives the graduate a very erroneous idea of the profession he is about entering. It cannot be too strongly insisted upon that the school cannot make an engineer, even if it crams him with all the book learning of technics. It can with perfect propriety give a young man a certificate that he has had a certain training, but he is entitled only to the term *engineer* when he has enough experience in manufactures, mines, or public works to qualify him at least as a capable assistant. The post-graduate system commends itself to my judgment as the wisest one, not only on account of moral maturity of character that a student of the average graduating age is expected to have, but also because the effect of a year or two of preliminary shop practice on the general run of boys would be fatal to habits of study. The exceptionally bright and ambitious youth would no doubt derive the immense advantage of the inductive system before referred to, and enter the school from the shop with a mind singularly well prepared for grasping the laws of scientific relations, but the average boy is not inclined to study, is not particularly ambitious, and in all probability would not reach the school at all. Once in the shop, he will in all probability conclude that he can get there all he wants to carry him through the world, he gradually ceases to study after hours, and so losing the habit, is satisfied, after his apprenticeship is finished, to take his stand in the broad phalanx of the rule-of-thumb members of the profession ; or, what is quite likely, avenues of earning wages will be opened to him, and he may deem it too much of a sacrifice to sever himself from such attractions.

Upon the second topic presented for our consideration, viz., the connection of experts with technical schools, I have long held positive views. It seems to me an anomaly in technical education to abandon its methods and direction to those who have either had no experience in technics, or experience in a very limited degree. If engineering is anything, it is a practical science ; it is the "art of directing the forces of nature to the service of man," and being such, it is the business of the schools undertaking to teach it, to show as far as possible how those forces are really controlled and applied, as well as their nature and laws of operation. This last the mathematical and physical professors can do very well, but the former can only be properly taught by the expert, familiar not only with principles but practice. If schools of law or of medicine cull their lecturers and teachers from those most skilled in the practice of their respective specialties, how much more ought our engineering schools to select for their chairs of applied science, men who have had experience as practicing engineers?

There are things, little things, perhaps, in themselves, but often keynotes to the comprehension of a subject, that the books cannot treat of, and the professor who depends upon what the books say for his practical teaching cannot appreciate their importance. The proper and feasible way, in my judgment, to incorporate the expert idea in any technical systems, is in the form of lectures as part of any given course, such lectures being confined purely to the engineering practice of, or the specialty of the lecturer, the lecturers not to be part of the resident faculty, but practicing engineers, able to devote a week or two in each term to an exposition of the methods and operations of practical engineering. This can be done, too, without trespassing upon all that is valuable in the theoretical portion of the technical course. As now administered too much time is given to abstract mathematics, and in some of the schools it is carried to such an extent, that the student, if he is not utterly befogged, becomes saturated with the idea that mathematics is the Alpha and Omega of the engineer's profession, and that he is the best engineer who is most expert in devising intricate formulæ and mathematical transformations. This is the natural result of what may be called the engineering of the schoolmaster, who tells us that the mental discipline and logical methods engendered by extreme theory are of immense importance to a young man desirous of practicing the profession of the engineer, and that the mathematical pathway is the surest road to eminence and success. Now, regarding the first claim, it elevates discipline at the expense of knowledge; as to the second, it is so far from being a sure road to engineering success as to be almost an educator of blunders, by giving a fictitious importance to the value of purely theoretical deductions. Engineering, having to do with natural facts and phenomena of ever-changing characters, must generalize knowledge from experiment and from the laboratory, and not from the hypothesis of mathematics. I do not wish to be understood as despising theory, but the conditions of practical life are such that to the average student, beyond a certain training in elementary principles, theoretical discipline is a waste of valuable time that could serviceably be spent in the accumulation of practical experiences and the observation of phenomena. In my preceding remarks I have not touched upon the matter of general culture, which I hold to be important to the success of the engineer, and I heartily indorse all that Professor Raymond has said upon the subject. I would say, in conclusion, that I look for any modification of the present system of technical education in the direction of a reduction of the amount of theoretical instruction, the introduction of a system of expert lecturers, a systematic study of experiments and their bearings, and the withholding of degrees until a student has qualified himself by a course of post-graduate experience. As to general culture, it should form part of the matriculation requirements, and not be attempted directly in the technical curriculum. The extent of this culture is to be as broad as possible, equal to that of the average college or high-school graduate.

Beyond the directions here indicated, I do not believe that the technical schools of the country will find it practicable to modify their systems and courses of instruction.

THE CHAIRMAN.—A gentleman, who I am sorry to say has been obliged to leave us to-day, Prof. Cook, the State Geologist of New Jersey, and the head of a very efficient though not very extensive school in Rutgers College, who began life as a practical engineer, and has been more or less in that work ever since, mentioned to me before leaving, his regret that he had not spoken last night, for the sake of bringing forward something that had engaged his attention for twenty years, and that seemed to me, as to him, to be one of the most immediately practical steps towards a needed reform. If Prof. Cook had remained, he would have been pleased to find in the paper just read, a forcible presentation of the point in which he was so much interested, that is, that the granting of degrees to civil and mining engineers should be made to depend upon their post-graduate course, and that the degree upon graduation from the school should be simply that of Bachelor of Science. That is almost exactly the idea Mr. Boller has brought forward. I will now call upon Mr. Coxe, of Drifton, Pa.

MR. ECKLEY B. COXE, *Member of the American Institute of Mining Engineers.*

MR. CHAIRMAN: I have been asked to give my views upon the

subject of Technical Education with special reference to mining, and although a member of the committee who prepared the programme for the speakers, I think I shall follow the very sensible course of most of the gentlemen who have preceded me, and ignore the aforesaid programme altogether.

One of the first questions that suggests itself when we begin to reflect upon this subject is (as I stated at the Washington meeting of the American Institute of Mining Engineers), "What is an engineer?" or "What do we wish to produce" when we submit a boy to the best system of technical education we are acquainted with? The tendency to-day is to make the ordinary workman a specialist, to teach him to do one thing well, or to work on one machine well, and then to keep him at that for the rest of his life. In old times a machinist, for example, worked at all sorts of jobs, and with every tool in the shop, and thus became practically familiar with all the details of the trade, and if he was a man of more than ordinary ability and industry, he might rise to be foreman, and afterwards director or engineer of the works. But a man who works all his life at one machine only, will undoubtedly become very skilful at that sort of labor, but will not be likely to acquire such knowledge as would fit him for the position of foreman. The fact that all the workmen are specialists, requires a higher grade of instruction and intelligence in the foreman. The workmen do not and cannot think for themselves; they must do exactly what is set before them, without any regard to what is to become of the work when done. Now the foreman should be a man well up in the theory and practice of the construction of machines, and he should have first-class theoretical training, and also first-class shop training upon all the machines and at all kinds of work; in other words, he should be an engineer, a man who understands theoretically and practically the laying out and carrying on of work; one who, when he has made the plan of a machine, a bridge, a furnace or a breaker, knows if it can be built or not, and how to direct the men who do the work. To obtain such men shop training is necessary, and the number that will be wanted is increasing. It seems to me that the standard of work is higher than formerly. Then the only requisite for a bridge or machine was, that it should be strong enough to do the work required, or to resist the strain to which it was to be subjected; now it must be constructed not only with a view to proper strength, and distribution of material, and adaptability to the work it is to perform, but also with due regard to its appearance. Great ugly masses of iron without architectural effect are no longer tolerated.

This requires a certain amount of æsthetic culture in the engineer; for that reason I fully concur in the opinions which have been expressed as to the importance of giving a liberal education to the student of engineering before he commences his technical studies. We all agree that a thorough general education should be given to a young man before he enters the shop, and the only question is, should shop work precede, accompany, or follow the theoretical studies?

My own opinion is, that a year's training in a machine shop, previous to entering upon his purely technical studies, is of the first importance to any young man about entering any branch of the engineering profession, provided he has had drilled into him, in his liberal education, enough of the fundamental laws of nature to prevent him from being affected by the shop superstitions, as I may call them; in other words, he should know that there is no effect without a cause—that when a man tells him (in referring, for example, to a machine) "*she works badly*," "*she takes fits of working*," etc., he must understand by it that some part of the machine is out of order, or not properly constructed. If the machine does not work right, it is because some man has not done or is not doing right.

I remember when I was about twelve or thirteen years old I tried to construct perpetual motion. It was very simple. I made a small water-wheel, and attached a very primitive pump to it. The pump was to raise the water to run the wheel. I was totally ignorant of the principle of the conservation of energy; if I had not been, I should not, of course, have attempted any such absurdity. Had I entered a shop in my state of mind at that time, I should have imbibed false ideas which it would have been difficult for me to free myself from afterwards. These remarks apply with more force to such complicated structures as furnaces. They are generally thought of and spoken of by the workmen as living beings, subjected to fits of good and ill-humor, instead of being simply large chemical apparatus.

I think it important for a young man to be thoroughly grounded in the principles of mechanics, physics, chemistry, and mathematics, so as to know what is possible, as well as what is impossible, before beginning practical work. I find in the practice of my profession that in many cases it is of very great use to me, to be able to say decidedly, "It is no use to try to do that; it is physically and mathematically impossible," and every engineer has suggestions made to him which are physically and mathematically impossible. I do not believe that the success of some of the most distinguished engineers can be adduced as a proof that one method of training is better than another because they were trained in a certain way. What we want is the best method for the greatest number of average minds. We all know that such men as Watt and Stephenson would become able engineers under any system of training.

There are others again of whom no system of education would ever make engineers. The point is, to produce the best marketable product with the available material. One of the gentlemen who preceded me said that a boy when he went into a shop might like the work, and prefer to remain and get good wages instead of going back to school. No doubt we will often meet with such cases, but I think the best thing that boy could do would be to stay in the shop, for by taking him away we might make a poor engineer and lose a good mechanic. I do not think the objects of our schools should be to turn out the greatest quantity of engineers.

I now come to the point upon which I was requested to speak, and I shall be brief. If a young man wants to study mining engineering, he can easily get a place about a mine where he can learn much that will be of use to him afterwards, without interfering in any way with the running of the works. Where there are corps engaged in surveying the mines, if he is willing to work day or night, where it is clean or where it is dirty, he will always be able to get employment as a chain carrier, etc., for there are always in the mining corps one or more men who are simple laborers, and a willing young man of intelligence could always take the place of such a person, and learn a very great deal, not only about the methods of surveying, but also about the way in which the works are laid out and carried on. If the young man has a certain amount of knowledge of drafting, trigonometry, etc., he can also make himself useful in plotting and calculating the surveys, and he would, of course, learn much more himself.

I think, however, as I said before, that at least a year's training in a machine shop is of the first importance for all students of engineering, mining engineering included. A thorough knowledge of machinery is of the greatest use to the mining engineer, whether he be located near to or far from machine shops.

I have always regretted that I did not work longer myself in a machine shop. There is no trade requiring such familiarity with accuracy of measurement. The intimate contact with the workmen gives the student a knowledge of their ideas, feelings, and ways of thinking, which will be of great benefit to him when he comes to have men under him.

If I were asked to advise a young man who had received a liberal education, and who wished to make himself as *useful* a mining engineer as possible, as to his course of study, I would tell him:

1st. Work a year in the machine shop, foundry, drawing-room, or blacksmith shop of a small shop doing general work in the mining region, before beginning your technical studies.

2d. While at the shop try to learn something about the mines in the neighborhood.

3d. During your time of study see such shops and works as you can, and spend one or two of your vacations in the corps of some mining engineer who makes surveys and does other mining engineering work.

4th. Study bookkeeping, and try to learn while in the corps, how the books and accounts are kept at the mines.

5th. When you graduate from the technical school, get a place, no matter what, at some mine, and if you are of the right kind of stuff, you will soon rise to be a really USEFUL mining engineer.

6th. Never despise *theory* or *practice;* both are equally important. Always try to know *how to do a thing,* and WHY.

7th. Try to make engineering a profession, not a trade—a thing to live *for,* as well as *by*—and do not let yourself get rusty even in those branches which do not seem to be of use to you in your business.

MR. WILLIAM P. SHINN, *Member of the American Society of Civil Engineers and of the American Institute of Mining Engineers.*

MR. CHAIRMAN: The special requisites in the education of the engineer, to which I shall refer, may not be considered "technical" by many, but their importance has been demonstrated to my mind in the course of my experience, and I call attention to them because they do not appear to be considered by professional educators as among the requisites, and because while other parts of this discussion will be committed to or assumed by persons far abler than myself, these particular branches of education are not so likely to receive the attention which, in my opinion, they merit.

I refer to, 1st, the science of accounts, and 2d, business forms and principles.

"Accounts" are, or should be, to the engineer, what history is to the statesman, the record of progress in a financial sense as history is in a political sense; and in many cases the chart indicating the shoals and quicksands to be avoided, the channels and roads to be followed.

"Business forms and principles" are as necessary to be understood by the engineer as by the lawyer, although they are not considered as forming part of the "technical" education of either.

Many engineers, civil engineers in particular, are so situated when in charge of important works as to be obliged to exercise functions both executive and administrative, and even sometimes and in some sense judicial. Like the captain of a ship, the engineer has frequently to be "a law unto himself."

Lack of knowledge of these forms and principles lead to many embarrassments and frequently to pecuniary losses. I have known an engineer of sufficient experience to have had charge of construction of an important division of railroad twenty or more miles in length, who, when the railroad company failed to pay its contractors and men, gave and accepted orders in favor of merchants, of which he kept no record, so that when the men, who were settled with before the merchants, got their pay, they were paid in several cases without deduction of the amounts of their orders, which orders having been accepted by an authorized agent of the company had to be recognized and settled, the amounts being a loss to the company.

The forms and general principles governing the law of contracts, ought to be understood in at least a general way by every engineer. His business is largely carried on through the medium of contracts, and the loose manner as to form, and indefiniteness as to understanding, in which agreements are often made by engineers, are the causes of many losses to, and much litigation with, their employers.

"We will make it right," is an expression very frequently used by engineers towards contractors in relation to claims for "extras," or alterations in plans, but with no definition of what is "right," it is scarcely to be wondered at that after the work is done, the parties can seldom agree as to the compensation. Such expressions are not businesslike; much less are they professional. Engineers should be men of precision; their technical education all teaches the value, nay, the *necessity*, of precision in measurements and in calculations, and it should alike extend to their commercial transactions, to their business expressions.

But it is in the matter of "accounts" that many engineers fail most completely. *More engineering works are financial than physical failures.* How rarely is it that the cost of a work of any magnitude falls within, or even approximates, the estimate made before it was commenced. A notable instance to the contrary was the Croton aqueduct in its original construction, the cost of which (approximating ten millions of dollars), fell within *less than one per cent.* of Mr. Jervis's original estimate. How frequently, on the other hand, could the projectors know the actual cost in advance, would they have declined to make the investment. So universal is this experience that it has brought actual and merited discredit upon the name of "engineer." Why is this the fact? Partly because engineers are not *precise,* but mainly because they do not keep "accounts" worthy of the name. If they and their predecessors had kept strict and rigid accounts, properly classified, of all expenditures upon similar works, they would have had records which would have been a guide for estimating the cost of future undertakings of the same character.

But the so-called "science of accounts" (for it is after all only a pseudo science) appears to be a "bugbear" to engineers generally, and yet nothing is more easily understood if the engineer will only bestow upon it that attention and give its principles that analysis which he so constantly is compelled to use in the line of his profession.

Early in my professional experience I heard an eminent civil engineer, then in charge of the construction of an important railway, say with satisfaction, if not with pride, "I have nothing to do with the finances or accounts." Yet he was the only man in the general management with any experience or technical education in regard to railroads. What wonder, then, that the road, estimated to cost about two million dollars, cost between four and five millions; or, that its accounts intrusted to unprofessional hands and without intelligent supervision, fell into confusion, and that the company after paying dividends for four or five years defaulted in its bonds in the next two years and had to be "reorganized?"

Another eminent railroad man, himself a graduate of West Point, used to declare that "accounts are an invention of the devil to cheat people with!"

The principle governing "bookkeeping," as it is technically called, is as simple as it is universal, and every engineer can and should understand it thoroughly. But accounts in general are not book-keeping in form, but rather in the form of records or abstracts, and it is in this form that they are most valuable to the engineer. In this form records can be kept which are not only valuable as showing where the money went, but they serve as beacons to warn the capitalist from too costly undertakings, and as guides to the engineer for estimating and constructing, and to some extent thus enable him to get at a measure of economy in the prosecution of his works. Such records, it is almost needless to say, are rare in this country. In an experience of over twenty-five years with public and private works, and an observation extending over many enterprises in the management of which I had no part, I have found such records to be the rarest exception.

Until this is reversed, until engineers are taught that they are responsible for economy in cost of construction, that economy cannot be attained without rigid accountability, that responsibility cannot be located without adequate accounts and records, and that it is the engineer's business to direct and supervise their keeping—until this condition of things is reached, we shall look in vain for that confidence in engineers' estimates and in the economy with which works have been constructed, to which the profession should be entitled.

PROF. THOMAS EGLESTON, *Member of the American Institute of Mining Engineers.*

MR. CHAIRMAN: It is certainly remarkable that among all speakers on the subject of Technical Education, there should have been but one opinion with regard to the necessity of previous culture before entering a technical school. It is all the more remarkable that the idea should be so prominent in the minds of so many engineers, when the practice seems to be to introduce young men to professional schools as rapidly as possible, without much previous culture. The highest possible previous culture seems to me to be an indispensable requisite. In the technical school the student has very little time to acquire anything but what strictly belongs to his education there. It is true that the amount of discipline which he gains from thorough training in the different branches of engineering will be of service to him, but he needs culture before he enters, for culture implies habits of thought, and no person needs these habits more than the student in the technical school. We are all of us creatures of habit, and it is therefore necessary that we should not only form habits of doing right, but also of thinking right. Such habits as these *must* be acquired before the profession is entered, and should be before the technical education is commenced. Once in practice, there is no profession in which the necessity to have a correct judgment, and to act on it quickly, to think, and to think rapidly, is more imperative. Accidents are constantly occurring about the mines and furnaces which require instantaneous decision, and therefore the exercise of immediate judgment. If the early habits have been right, the engineer will think right from habit. If he has been accurate as a student, he will probably be accurate as a man, and for that reason he should have not only the highest possible preliminary education before he enters the technical school, but the very best after he enters it, and this education should include the higher mathematics, as being one of the essential means of learning to think accurately. I have been censured in my professional chair for demanding and exacting of the student the attainment of a thorough knowledge of the higher mathematics, but I have done so on the ground, not that he will in ordinary cases be called upon in practice to work out difficult problems, but that the mental discipline and the correct methods of reasoning which are given by thorough familiarity with mathematics, imply the likelihood of correct methods of reasoning and accurate judgment in other things.

We are very apt to think that the man who has learned the most is therefore the best educated and most learned man. It is sometimes said of men that they have forgotten more than their neighbors ever knew, as if this was a credit to the one or a discredit to the other. The best educated man is not necessarily the one who has learned the most, but the one who has his knowledge thoroughly classified, not overcharged with detail, but the detail so arranged that he knows where to find it when he wants it. The study of mathematics, when properly pursued, with other branches, gives such power of classification, that, even though it was to be little used as a sign language, it would be a valuable part of technical education for this reason alone.

Besides mathematics he should be able to use with a certain degree of facility at least two modern languages besides his own. If he can learn but one, it should be German, as being the one in which the greatest number of engineering works are now published by men willing to devote their lives to specialties, and in which an inexhaustible fund of information exists. If he can learn two, the second should be French, as being the key to what was formerly done, and to what will probably be done again in engineering science. The acquisition of at least one modern language is almost an indispensable requisite to success. More important than either of these is the sign language, drawing. He should become as far as possible an artist in both free-hand and mechanical drawing, for, as a master of his pencil, he will be able to communicate his own ideas with his fingers, and to understand those of others which could not perhaps be readily expressed in words. I do not think that sufficient importance has been given to drawing as a means of expressing thought in our system of education. There is too much drawing done from models which is absolutely necessary to the beginner, but we do not require the student often enough to express *his own thoughts* by a drawing, just as he would be called on to express them in words. This could be accomplished in the same time that is now devoted to it, but would require a different system of instruction.

Dr. Raymond remarked at the Washington meeting that we had taken from the German university system everything except the time it takes, which is lamentably true. In this age and in this exciting climate we live in a hurry, we work in a hurry, and die in a hurry. We commence too young, enter our preparatory schools and colleges with too little preliminary training, and before we are physically and mentally able to comprehend what we are taught. Our high standard then, if premature, becomes simply a forcing system, and until we can do away with it, we make engineers on the principle of the survival of the fittest. It is no wonder that so many young men give out before they leave the intellectual hothouse, or occupy a medium position after they have left it.

It is undoubtedly true that all our technical schools would be in a much better condition if a part of the school course was thrown into the requirements for admission, and the age required for entrance advanced. I think that every faculty would be unanimous in this respect, and would consider that there would be a great gain in detaching from the course proper all of what must be considered purely preparatory work, and supplement it with advanced studies. This would have the effect to diminish the number of students, but to raise the grade of instruction. The opposition to this course would come neither from the young men nor the parents, but from the boards of trustees, who are apt to consider that the number of students in an institution is the chief if not the only real criterion of its success, while those of us who are engaged in education know perfectly well that numbers, instead of being necessarily a sign of power, may be a decided indication of weakness, though not necessarily so. Every faculty would agree with me, I think, in saying that it is far better to graduate one first-class man a year, than to graduate twenty indifferent ones. Unfortunately, there are but few institutions with endowments sufficient to feel justified in pursuing such a method, and fewer still who could divest themselves of the idea that numbers are in some way the best evidence of success. There may be no lack of ability, but to make up the average we often dilute that of a high order in order to keep up the lower grade. It is no just criterion to look at the expense per capita if five men of a high order were turned out where fifty of a lower grade might be. As a matter of arithmetic the expense is greater for the five, but as a matter of political economy it is really less.

The training of the schools should be such as to teach the young engineer how to acquire not only judgment, but as far as possible an independent one. This I believe to be one of the distinguishing characteristics of American technical education, while in some countries in Europe exactly the contrary is true, and the best reason that a young engineer can give is, frequently, "My chief told me so." No profession requires more judgment than that of a mining engineer. We have constantly something to do, and that something involves either life, present or future safety, present or future capital. The same problem rarely ever presents itself twice in the same way, and if the preliminary training has been such that we have acquired correct habits of thought, we probably shall carry the same into our profession, or at least be able to find out and correct errors before they have done harm. Mere facts are of very little use to the engineer of mines. We want not so much the facts, as the conclusions which are to be drawn from them, and since the data are rarely ever the same, we must act upon conclusions.

Dr. Wedding has hit the keynote of the whole subject. He wants his students to be able to settle at about thirty, and our students want to settle at about twenty-two. It is very true that, according to the old proverb, late marriages make early orphans, but it is also true that the average man has not acquired, or cannot have acquired, at the age of twenty-two or three, judgment or experience sufficient to justify capitalists in putting him in such positions as would enable him to settle and marry.

Many young men are in a hurry to complete their education, owing to lack of necessary means to prolong it, and the necessity of commencing or being compelled to earn their living in some other way than the occupation of their choice, is often the reason why they commence prematurely. They feel, and with some show of right, that, in order to be established in their profession by middle-life, they must commence early. While this desire to hurry has been in some respects a disadvantage, it has been in others a real benefit to us. As a nation we are now just one hundred years old. The political necessities of the case demanded that our organization as a nation should be made in a hurry, and since we started in that hurry, what have we accomplished? There are but few nations in Europe who have as much to show in the same period. We have had no time to deliberate, for in many cases if we had taken time to deliberate, we should have taken still further time, and not made a move. Ten years ago we crossed this continent in stage coaches, a trip which cost three months after leaving the rail. We have recently accomplished that same distance by rail in eighty-eight hours. I had the good fortune to be present at the commencement of that road, and to see the first laying of one mile of track in a day. They were indeed in a hurry, but it was an organized and therefore a successful hurry. I think it may be said that where we fail, it is for the want of organization of this hurry. It may be said that, as a result of this method, many things that we have accomplished could have been better done, to which we may reply, that the necessities of the case have almost always been such, that, if what has been done had not been accomplished in this way, it would never have been accomplished at all, and therefore, even by imperfectly doing what might have been better done, if the element of time had not been that of which we had the least, we have not only benefited the present but the future.

It has been urged by many persons as a reason why the student should enter early upon his professional career, that education unfits a man for work if continued after a certain period of time, to which we can only reply that any system of education which unfits a man for work, or leads a man to despise or even to look down upon it, is a system which is fundamentally wrong. Labor is one of the best gifts of God. The man who is afraid to dirty his hands has mistaken his profession, and should be made to feel that he is out of his place and mistaken his calling. I had an apt illustration of this in my early professional experience, when four persons were selected from a prize examination to enter upon some difficult work requiring ability, judgment, and some little dirty work.

Their expenses were to be paid, but there was no other pecuniary reward; the honor of solving, or being selected to solve, the problem being considered by us all as a sufficient reward. Their intellectual merit was about equal; but when it came to the question of dirtying their hands, one of them presented himself in gloves, patent-leather shoes, and declined to put his hands into the dirt. As I look back after ten years upon the course of these men, I find that the one who was afraid to dirty his hands became disappointed, and left the profession shortly after he entered it, as it was evident that he had mistaken his vocation. The other three occupy distinguished positions in the profession.

One serious difficulty with students or young graduates is that when they enter the works they feel themselves to be the superiors of the workmen, and they are apt to patronize them, which the workmen resent. No one knows better than the workman that the student directly from the technical school, though he has the title, is rarely ever an engineer. He has simply acquired the a b c, and is ready to commence to learn to be an engineer. He has to learn what the workman is doing, and why he does it. The intellectual, and perhaps the social, superiority, the workman might be willing to admit, but he knows that the young man is entirely incapable of doing what he can do with ease, and therefore feels himself his superior, and he is right. We never cease to admire the skill of a good workman. He is a real artist who has learned from long experience to seize with the nicety of exactness the instant both of the commencement and end of reactions and changes which are entirely inappreciable to any but his educated eye. He has learned this from years of toil, and labor, and though he might not be able to explain them, he can always perform them with precision, while the student, if he has learned the reasons, is not able either to do the manual work or to distinguish between the different parts of a process. Any young man fresh from the schools who looks down on the working-man has a totally false idea of his own and the working-man's place. The two occupy a totally different position. The student can learn, and learn rapidly, all that the workman knows, and when this has been done, no one is more ready than the workmen themselves to acknowledge his superiority, and to sit at the feet of the very man whom but a short time before they looked down upon as an inferior. It is therefore a serious mistake for young men who wish to study to place such an antagonism between themselves and their teachers for the time being, as to prevent their acquiring the information by placing an unnecessary barrier between the workmen and themselves.

There is one thing which it seems to me eminently desirable that we should introduce into our instruction, and that is the economics of the profession. It has been my own habit ever since I first commenced to teach to consider the cost of a process as the subject about which the most detail should be given, and to which the greater part of the instruction should be devoted. I have therefore made this the framework of my instruction, and have considered the theory of the process as the means of illustrating why the cost should be in certain cases greater or less. It has been my habit to define metallurgy as the art of making money out of ores, and to announce at the very outset of the course, that an ore was a material sufficiently pure and rich to allow of the metal being extracted from it with profit. By insisting that an ore is not an abstract thing, but that it represents labor and the value of the material used to get it, it becomes in the minds of the young men a concrete thing. They see at once that its value depends upon its being worth more than is spent upon it, and that if it is only worth what it costs to get it, there is no profit in working it. By introducing at every step the cost of that part of the process in labor, materials, tools, wear and tear, interest on capital, etc., and by familiarizing them with samples of schedules for keeping tally, and account books for keeping the costs, they become thoroughly familiarized with the idea that a process is good or bad according as the balance is on the right or wrong side of the ledger, no matter how tempting it may look on paper. They thus learn that an engineer who neglects to know how and when he is gaining or losing money is to be blamed. In order to still further keep the commercial bearing before them, as well as the theoretical, and to teach them how to distinguish between the right and wrong side of the ledger, they are required before they graduate to execute a project which is given them in a strictly commercial sense, and which they must execute in both a theoretical and a commercial manner, by giving the theoretical reasons for its adoption, and estimates giving the cost in detail not only of the constructions, machines, etc., but of every step of the work in detail. It is a very rare thing that such a project worked out by a student can be executed, but this is not the design. The object is to keep them familiar in every stage with the fact that no work can be done, no time or labor employed, and no material used, without involving capital and the interest upon it, wear and tear of machinery and tools, and that they are, therefore, to look upon the scientific exposition of their profession not as a mere matter of theory alone, but as the combination of both theory and practice, which requires the scientific union of both, which can only be had in men educated in this way. Men so trained, brought into metallurgical works, have proved themselves to be worth a salary as they enter, and have shown themselves capable of rising in their profession with a minimum amount of cost to their employers.

In this age every man expects and is expected to make money. Every young man commencing has always in view a fortune, more or less large, amassed in a time more or less short. The first question of the father when he chooses an employment for his son, or his son selects one for himself, is, Will this or that profession enable him to become a rich man? No man expects to be one of the thousands of the unsuccessful, yet there are times in life, and there are commercial conditions, when losses are inevitable, and by making these losses an object of thought, and keeping them in the minds of young men, teaching them that certain processes and principles, if placed in practice under certain conditions, must inevitably bring about a loss when under different conditions they might yield a profit, I think we shall have accomplished an important point, by familiarizing men with the idea of loss. If the incipient engineer is taught that there are times, like the present, for instance, when the dominant question is, not how much profit we can make, but how little loss, we shall get rid to some extent of this idea of gain, and add an impediment to the idea being received that the generality of mankind can make money by other means than by working for it, and not always then.

The graduate must also enter the profession thoroughly convinced that he is not an engineer, though he may have the title, but that he has in his education the means of becoming one.

Above all things his education should teach the young engineer that honesty is not only the best policy, but an absolutely indispensable criterion to success. Every young man gains his experience more or less at the expense of his employer. What is produced is in his charge. It represents in a great many cases, besides the capital engaged in it, the wear and tear of human life. If he is strictly honest with himself he will understand that upon his endeavors are based the happiness and prosperity, and in some cases even the lives, of many a family, and often the interests and fortunes of his employers. It requires in a great many cases more than ordinary courage to say he does not know the answer to a question, but will study it, and not try to improvise, when he should say that he requires to study. It is one of the mistakes of both employers and engineers to think that, because a man has a diploma, he must therefore be an encyclopedia, and the anomaly of the thing is, that this knowledge is demanded only of young men just entering the profession. No man would ask a distinguished engineer long in practice to solve a difficult problem on the spur of the moment. The engineer needs time to study and consider, and the employer demands that he shall take that time, but although in the case of the student the same interests are not involved in so great a degree, they are in a less, and the rule in a great many cases is not made to apply, but the young men are asked and expected to answer questions at once which can only be answered conscientiously after a more or less long study.

Every man who is an earnest worker makes mistakes. The man who makes no mistakes is free from them because he does no work. The wise engineer will always endeavor to find out what these mistakes are, and correct them before they have done any harm.

There can be no doubt that theory and practice should go hand in hand. It seems to me, judging from my own experience, and that of a large number of engineers, both in this country and abroad, with whose experience I am familiar, that the practical course, which we all consider to be absolutely indispensable, should follow the theory, that while the theory is being acquired, the student should see, as far as possible, the processes or machines which are being described, and the theory of which he is learning developed in practice, so that he may have some idea of what the process or machine is, without perhaps being familiar with it in all its details, then when the theoretical knowledge and the power to think correctly has been acquired, the practical course should commence. I do not understand by this practical course that the student should undertake, or should be expected to undertake, the work of a full shift. Very few students have the physique to do it, but they can work a turn, or a part of a turn, a phase, or a part of a phase, of an operation or process, so that with three or two, or perhaps no hours at all of manual labor a day, they may in a time more or less short become perfectly familiar with what they have learned theoretically. If there is not sufficient physique, a great deal can be accomplished by watching the workmen, and hearing them discuss, and learning to sift what is really good from what is useless in their conversation.

More is often learned in this way by observation than by actual work, and though there is no doubt a great influence to be gained over him by the muscular power of doing all that the workman does, and in setting him right when he exerts that power to bad advantage, it is doubtful whether the result is really worth the time which the young engineer must take to achieve it. We should not lose sight of the fact that what is needed in this case is not brute force to work, but the acquisition of a correct judgment, which is gained mostly by ear and eye education. Theory, without this ear, eye, and perhaps hand practice, except in cases of real genius, can never be expected to produce good results, and practice without theory is very much like endeavoring to do good work with poor tools. An artist can do it, and none but an artist can, but the world at large cannot depend upon artists, for genius is apt to be erratic and unpractical, and while it is undoubtedly true that the larger part of the workmen who execute the very intricate and elaborate practice of our best engineering and metallurgical works have very little idea of the theory of the process upon which their practice is based, it is also true that many of them have become practical artists, and have almost made their trade a profession, but they have taken years to learn by routine what an educated man will acquire in a short time. No one knows better than a man of education that he cannot always believe what he sees. He has learned it in physics, from the phenomena of persistent vision, and no one knows better than he does that he must learn to distinguish between what really is and what he thinks he sees, while an uneducated man thinks that if he cannot believe his eyes there is nothing to be believed.

There is a phase of technical education which I have never seen carried out anywhere except in Krupp's works, but which I believe to be practiced in some of the large works of France, and which applies only to those works where there are a large number of engineers associated together. At these works a council of all the engineers of the establishment under the president, Mr. Krupp, is held at stated times. The problems to be worked out are there discussed, and the executive officers for their solution appointed. Besides this the younger members of the corps meet frequently to discuss problems. Whenever anything new comes up which requires experiment, a certain amount of funds is devoted to that purpose by the works, and the experiments are made under the immediate supervision of those whose business it is to carry them out, but with the active co-operation of all the engineers of the works. It is not wonderful that with such a technical school as this, established in the works, the works themselves and their proprietors have achieved such a reputation. I consider this establishment a permanent and irrefutable argument against the opposition of owners of works in general to allow young men to take practical courses in their works. We have besides in our own midst another answer to it, which is the general practice of the Bessemer steel works of this country. I am quite certain that but for this wise policy the American Bessemer plant could never have increased its product to such an extraordinary extent. I can understand perfectly that owners of works should object, and I should object as well as they, to young men standing idle round the works, often in the way of the workmen, asking questions that cannot be answered, or that would require long study to answer, with no intention of making any use of it except to perhaps retail the information at secondhand; but this is not the case we have in view.

One of the objects of this discussion is to ascertain whether in the minds of experts a practical course before or after graduation is desirable, and whether in the interests of the owners of works it is practicable. My own opinion, based upon nearly twenty years' experience, is that it is more desirable afterwards, and that it is really the interest of the works to have such a course carried on; but carried on in such a way that the young men shall study, and be required to show some fruits of their study, instead of being allowed to go around in the works at their will, and ply the engineers or workmen with questions. Such a practical course as this could be carried on with no trouble to the works under the supervision of the engineers, and I can readily conceive might become a source of real pleasure to the engineers in charge, and of profit to the works, for no student not anxious to learn would ever think of undergoing such fatigue and labor as would be necessary to gain the information sought. The engineer in charge would be surrounded by young men eager to learn, and ready to attempt the solution of any problem presented. The only inconvenience I can conceive would be that the engineers in charge of the young men, would be obliged to be extra-cautious in advancing theories or in making statements. It is certain that such a system, if adopted, would very soon train a class of engineers which are not now to be had, except in some few rare instances, like that of our honored President, who, by practice, has become so familiar with his profession as to feel in his own body when anything was wrong, or likely to be wrong. I am quite certain that if the managers of works can be brought to see how much for their interest it is to have educated young men, eager to learn, around their works, we should soon have upon the stage a class of engineers who would, by careful study of theory and practice, in the course of a short time suggest such changes as would not only diminish the cost of production, but increase the product. If such places were made the reward for high attainments in the technical schools, and were coupled with a system resembling that of the English university fellowship, we should soon have young men studying in the works at home and abroad, all those intricate problems which the necessities of the case force individual countries to develop. It would not be a great many years before the reflex action of such study would be felt by the manufacturers themselves. There is a groping more or less in the dark for such a system in literature, and to a very limited extent in science, in some of our colleges; but it fails of any great influence because as yet it is aimless, beyond the attainment of a certain amount of information, and because the privileges of the provision last too short a time. I have been endeavoring to introduce it into professional engineering for some years with great success, but without much financial support. Some few able young men of wealthy parents have followed it under my direction most successfully, but no system can be regarded as a real benefit where the application of it is based on the accident of wealth. We need endowed positions like the fellowships in the English universities, which will allow the student who has won the position to pursue, either at home or abroad, a system of post-graduate study of those branches which the circumstances of the case has made specialties in certain countries, or to undertake under competent direction, investigations which require time, and which could not ordinarily be undertaken except under some such conditions. There are multitudes of such investigations in the profession which are waiting, because the necessity of making them *now* is not apparent to the manufacturer, or because no one works could undertake the expense. In support of the proposition I can point not only to what the English fellowship system has done for literature, theology, and philosophy, but also to what the Swedish iron comptoir has done for metallurgical science, not only for Sweden but for the world. It is very doubtful whether but for this institution the Bessemer and many other processes would not have been declared failures. It has supported and maintained investigation, theoretical and practical, which never could have been commenced but for the intelligent foresight of the managers of that body.

In this country it could not be expected that any such endowment as that of the Swedish institution could be accumulated in a short time, but our technical schools might easily establish a post-graduate fellowship system, and it would not be very long under this influence before we should cease to hear of what cannot be done, and regard every question as a problem to be solved, with, for the time being, a plus or minus sign, which is susceptible of change under varying commercial conditions.

MR. ROBERT W. HUNT, *Member of the American Institute of Mining Engineers.*

MR. CHAIRMAN: As has been observed by so many of the gentlemen who have preceded me, there has been a striking unanimity of opinion as to the necessity of a good general culture, as well as a practical education. All have seemed to agree on these two points. The only difference of opinion so far in this discussion, as I understand it, is as to *when* that practical education should be obtained.

I think we are all governed in our judgments by personal experience, and it is, perhaps, for this reason that I have a strong predilection for placing the practical education before the thoroughly scientific one. Granting decidedly that, first of all, general culture is very necessary, I believe that, so far as metallurgical engineering is concerned, after the general culture should come the practical, then the purely technical education. I cannot judge for the other branches of the profession; I only speak of that of which I have some knowledge, and in which I have obtained some experience. I spent some four or five years in practical iron manufacture before entering the laboratory for a technical course, and the practical knowledge thereby acquired gave me a standing in advance of other students. I knew what it was that I was seeking to investigate, and it seemed to me that I arrived at some points quicker than other young gentlemen around me who were possessed of larger natural ability. When it devolved upon me to assume active duties and responsibilities, I certainly was better qualified than if my ideas had first been moulded by theory.

I agree with Prof. Egleston, that a boy cannot be expected to work as hard and produce the same results as a person of mature years—"cannot be expected to work a full shift in any branch of labor." It would be wrong for him to try to do so, for were he to exert himself to the full extent of his physical powers, his mind would not be in a state to investigate the theory of what he had been doing.

But there is one point which we seem to be overlooking. *Where* is the student to find these practical schools, whether they come before or after the technical schools? I freely confess that, as a rule, proprietors of manufacturing establishments do not like to have students in their works. While I believe this to be a mistake in other branches, I know it is a mistake so far as Bessemer works are concerned. If the plan suggested during the first discussion of this subject is to be carried out, that is, if the students are placed in charge of a competent teacher, who is responsible to the manager of the works, the establishment will be benefited. Owners will, I think, soon begin to realize this, and thus schools will be furnished. I hold that, as the student will report to his teacher what he sees and hears, coming to him for explanation; and that, as the teacher reports to the manager, the works will have just that many more overseers. The workmen will realize this, and so fewer departures from the desired practice will occur. The teacher will fully understand the policy of the management. He will, so far as necessary, explain this to his pupils, as well as the details of the works. While working with the men, students will observe all that occurs, and will ask and answer questions, and, while they act as a kind of moral check on the men, the men will be induced to think, and as soon as a man gets to thinking he is started in the right direction. Then, on the other hand, while the student is learning how to use his own hands, he also learns how much one pair of hands can do, thus laying one of the strongest foundations of success as a manager, viz., an accurate knowledge of how much to expect from labor; and his own physical man will be developed far better than by boating or other such exercises.

If this practical study is left until after the young man finishes his technical course, the danger is that neither he nor his parents will be willing in most cases that further time shall be spent on his education. The feeling will be, "Oh, the practice will come." So it will, but at whose expense? And it will not come in time to guide his theoretical studies.

No manufacturing process has made such wonderful progress as the Bessemer, and its gigantic strides in this country have come from breaking loose from old practices. This result has been to a large extent brought about by introducing educated men into the works, thereby creating an atmosphere of progress, a willingness and ability to get out of the old ruts. They are men who possess, beside a general knowledge of metallurgy, a thorough knowledge of the special branch; and they also have around them intelligent workmen.

I will cite the case of one Bessemer works, the one that is now turning out the largest product in the world. Their manager is a graduate of the Rensselaer Polytechnic Institute. Before assuming any responsibilities after his graduation, he entered upon a regular routine of work of some years' duration at the Troy Steel Works. After this experience he entered upon the management of the works with which he is now connected, and his career has been singularly successful. I doubt if this result would have been reached if he had taken a department even, of the management, immediately upon leaving the scientific school, as so many graduates have attempted or desired to do. So, while preferring that the practical course should precede graduation, we find good results from the other plan. As to the importance of both the practical and the theoretical education before the student ventures to practice, we all seem to agree.

PROF. C. O. THOMPSON, *Member of the American Institute of Mining Engineers.*

MR. PRESIDENT: I deem it a considerable privilege that the first time I have the honor of addressing the members of this Institute, the subject should be Education. For many years it has seemed to me that the future of all the industrial arts is conditioned upon the careful and philosophical unfolding of a system of technical education. What little I may contribute to this discussion, which has been so ably maintained, shall be to this precise point. And perhaps it will be useful to relate as briefly as possible, the results of an eight years' experiment in technical education in the city of Worcester, Massachusetts. In this school the different ideas which have been broached during this meeting have been subjected to a practical test, and a recital of the results will accord with the spirit of this discussion.

It is somewhat natural that a school of technology should be established in the county of Worcester. More than two hundred years ago the first school of this sort in England was founded by the Marquis of old Worcester. The population of its modern namesake is one-half of one per cent. of the population of the United States, while one per cent. of *all inventions* made in the United States is accredited to citizens of this county.

It is an interesting fact also, that a citizen of Massachusetts, Dr. Bigelow, is entitled to the honor of introducing the word *technology;*

at least no earlier reference to its use is made in Worcester's Dictionary. Indeed the geographical position of Massachusetts, and her dependence upon the prosperity of her varied arts and industries, renders technology, in its broader and its narrower import, a subject of great interest to her citizens.

Eight years ago I began to study this subject and to incorporate the results in an actual course of training for boys who are to be engineers. In organizing this school we had little light to steer by. All experiments hitherto made in this country, at least to combine manual labor in any form, have not been largely successful. Hence we have been compelled to move very slowly.

The Free Institute was founded by John Boynton, of Templeton, a tinsmith, who desired that apprentices in the mechanic arts should have a more intelligent preparation for their business. This endowment was largely increased by Stephen Salisbury, of Worcester, a Harvard alumnus, who desired that theory and practice should go hand in hand, and by Ichabod Washburn, founder of the Washburn and Moen Manufacturing Company, who merged in this scheme a cherished plan of his own for the education of mechanics' apprentices. Citizens of Worcester contributed money for the buildings and the State made a grant of $50,000. The whole amount of the plant is $600,000; the available annual income $25,000. The school is free to Worcester County, substantially so to the State of Massachusetts, and open to all qualified applicants.

The theory of the Institute is that boys who have the best training afforded by our common schools may enter, not younger than sixteen, upon a course of study which shall give them a good education based upon the mathematics, modern languages, and physical sciences, and such a knowledge of some form of handicraft or industrial art as will enable them to earn a livelihood immediately after graduating. The conditions of admission are fixed by the actual results of common school instruction rather than upon what ought to be expected of it. For instance, we find few boys who really understand algebra well enough to use it as an instrument of research. Many have been "through it," as the phrase is, but hesitate before such problems as, given $\frac{a^n - b^n}{a - b}$, required to show that a perfect division is possible, or given $t = \frac{a}{m - n}$, required to show the significance of each possible value of t, and how its value is affected by changes in the other term. We find that the only sure and safe way is to teach algebra from the beginning. A knowledge of arithmetic, geography, history, and English grammar we take for granted after examination. The course of study for all students then proceeds, for forty-two weeks in a year, for three years, in mathematics, through geometry, general and descriptive geometry, and the calculus, and blends with the course in physics and elementary mechanics, the careful reading of Rankine's *Applied Mechanics*. Synchronous with this is a course of free-hand drawing, mechanical drawing, physics and chemistry, and language—the English and either French or German. Lectures in geology are given, and the course in chemistry includes determinative mineralogy. So much is taught to every student in a three years' course.

Now ten hours a week, and eight hours a day for the month of July, each student practices according as he is to be a mechanic, a civil engineer, a chemist, or a designer in the workshop, the field, the laboratory or the drawing-room. But since the last named forms of practice do not differ essentially from the same work elsewhere, I will confine my remarks to the first.

The Washburn machine shop, part of the Institute, is a thoroughly equipped manufacturing establishment. Boys who have had no mechanical training enter it in February and spend most of their time there as apprentices till July, so that our mechanics' course is a half year longer than the others, for absolute beginners. There are five principles embodied in this shop:

1. *Construction* must vitalize and guide all *Instruction* in practical mechanism. If a boy turns a wooden rod it must form part of some structure, and if not found good enough, must be reproduced till it is good enough. Whenever two pieces of wood or of iron are joined, the resulting piece must be subjected to the absolute test of accuracy which is imposed by the highest standards of manufactures. In this particular our plan differs from that of the admirable schools at Moscow and St. Petersburg, where separate attention is paid to construction; but for American boys I am sure there is good reason for us to adhere, for the present at least, to the synthetic method. I was unable to get access to these Russian schools in 1868, when examining European methods, and could take advantage of their interesting results only at the Vienna Exposition. Hence the methods are fairly and simultaneously under experiment.

2. Practice must form a part of every week's work, in two periods of five hours each.

3. Students must be taught by the most expert mechanics, and

use the best possible tools and machinery. Nothing is too good for a boy.

4. Students must not expect nor receive any immediate pecuniary return for their work. It is impossible that the best men shall be employed as instructors, the best machinery made and used, all subjected to business conditions, and a large number of apprentices taught in small divisions, with any other return to the student than the advantage of learning.

5. Since instruction and not construction is the main motive, each student is advanced in the grade and quality of his work as fast as his attainment permits.

The shop is thoroughly equipped with tools and machinery for working wood and iron. The apprentices, that is, those who enter in February, work wholly in the wood-room during shop hours, and have training in drawing ten hours a week. From the beginning of the junior year the students practice in the iron-room, learn the management of the engine, and make working drawings of machines for use in the shop. We often receive at the beginning of junior years young men who have already worked in a machine-shop one, two, and even six years. These students are always numerous enough to enable us to compare the relative advantage to a student of preliminary shop practice.

The shop is under the direction of a superintendent, a graduate of the Chandler school, with six years of shop practice behind it, and two foremen, one in wood work and one in iron. The force of journeymen varies from six to twenty according to the condition of business. Certain standard machinists' tools are manufactured, viz.: An engine lathe (described by Prof. Thurston in Johnson's *Cyclopedia,* vol. 2, Art. Lathe), a speed lathe and an emery grinder. We also make a draughtsman's stand and sets of models used in teaching drawing. These we have brought here to the Exhibition and offer for examination in Machinery Hall, in order that you may see that this Institute is not a school with a shop attachment. We wish to find out in this most desirable and satisfying way, whether we are really giving our students the best models of workmanship, and illustrating the best American practice. I may also say here that the whole scheme and plan of the Free Institute are on exhibition in the east gallery of the Main Building, and the valuable examination and criticism of the members of these Institutes of Engineers are much desired.

To return a moment to the shop. It must be stated that in addition to its standard manufactures, contract work is done, such as the erection of hydraulic elevators, etc., with the intent to give increased variety to the work. Some of the more palpable difficulties in the business management of the shop are met by considering that the whole plant is a gift and free from taxation; that the early work of beginners does not involve a great waste of stock, and that, although students enjoy the advantage of doing all kinds of work which are done in the shop, whatever is actually *sold* must satisfy a rigid standard of excellence. I suppose we sell about $18,000 worth annually. The profit on these sales together with the income of the shop endowment fund of $50,000, enables us to meet all expenses. Or, if the total annual expenditure in the shop be divided by the number of students who work in it, the per capita is about $120 a year.

Now, I hope I have sufficiently emphasized the point that everything a boy makes in our shop is done with reference to some use. He does not make joints merely as such, or file away a piece of iron merely to learn the use of the file. He learns to make neat joints, but they are parts of a box or of a model. He learns to file, but the piece he files must fit into its place in the machine. So we try to make construction direct, and to temper and strengthen instruction. There is a certain proper and logical order of parts in the training of a mechanic, and this is not lost sight of. In short, the graduate of our school, with the practice I have outlined, we believe to be able to hold his own in open competition with the apprentice who has worked, under ordinary circumstances of apprenticeships, three full years; experience so far confirms this view. Here, as always when speaking of results, I desire to speak with great diffidence. The results are those of six classes and eight years. They are alluded to as grounds of our willingness to proceed with further experiments in the same direction.

Mr. President, I have gone on in this rambling way talking of *our* work, and I fear the story is wearisome. [Cries of, "Go on."] It seems to me that there are three elements in all sound technological training,—handicraft, technical training, and culture. There has been a surprising and gratifying unanimity of opinion on this point among the gentlemen who have spoken. The only question is how these elements shall be combined.

What is the object of technological schools? It is to produce the best possible raw material for skilled workmen. No school can impart the skill and good judgment which come alone from experience, and without which no man is a competent engineer.

Another question with which a school has little to do is whether a youth shall be a superintendent or not. I take it that the Almighty makes superintendents. The precise kind of talent which enter into their characters is strictly an endowment.

In this discussion some have held that the order should be handicraft, technics, culture; others culture, technics, handicraft; and others would arrange in other ways. But there is one objection to all these sandwiching methods. Practically we cannot hold our young men in training till twenty-five. They will go at twenty-one or twenty-two. The period of sharp acquisitiveness, the most precious part of school life, lies between sixteen and twenty-one. Now, whichever part of a boy's triune discipline for an engineering life is allowed to usurp that period to the exclusion of the others, that will be the dominant force in his after-life. If culture, then practice will suffer; if practice, culture will suffer. Either part will be, as it were, attached to, or subordinated to, the one which "rules the favored hour." Hence it seems to me that all possible culture should be secured before a student begins his technological course, and that it should be looked to ever after. It must not be forgotten that culture is a result, or rather a growth. All we can do is to prepare the soil. The plant will assuredly grow. Perhaps, too, the best and only useful culture is to be looked for in the life for which any school training prepares a man; for I take it, we are not now speaking merely of the cultivation of the æsthetic part of man, but of that discipline of the judgment, awakening of the imagination, sharpening of perception, repression of conceit, and elevation of motive, which constitute a serviceable and efficient man of refined taste and unquestionable integrity and courage.

Let us secure as large a foundation as possible in general knowledge before the beginning of the technical course, and not lose sight of the bearing and relations of all knowledge during this course. But let us blend technics and handicraft in the technological course. The drift of this discussion has been unmistakably towards the affirmation that the technologist of the future is to be the educated workman. It is to the man whose own hands can execute, if need be, the behests of his brain, that the great engineering works of the future are to be intrusted. Engineering, so happily defined by the retiring President as "the arts of production and construction," including mechanical, civil, mining, and chemical branches, more and more condenses into mechanics. Indeed, all branches of engineering seem to react upon mechanics, forming compounds like different acids upon a common base. We are coming to think that, if a man is to be a civil engineer, he had better begin by being a mechanic. If he is to be a mining engineer, he had better begin by being a mechanic. If he is to be a chemical engineer, he had better begin by being a mechanic. This is true, at least, of all study of *applied science.*

Now, as to the amount of preliminary culture, it is desirable that at least what is included in fitting for college should be secured. I do not think graduates of college in general will be drawn to technical pursuits. The whole drift of the college is averse to them. Few boys are so powerfully polarized as the sons of the honored member who spoke last evening. What might be very easy for Mr. Sellers would be very difficult for a father in other walks. In short, it seems to be the best available method for the average boy to fit him for college, then send him through a technical course in which handicraft shall find a place; then let him enter some manufacturing or engineering works, and see what it all means.

There are a few other points in the course at Worcester which I will mention, as throwing light upon some parts of this discussion. Whoever takes pains to read the examination papers, and consult the returns on them in the Exhibition, will see that the standard of scholarship is not suffered to fall. It is no part of our plan that the workshop instruction shall invade the province of intellectual preparation for an active life. If our graduates suffer in consequence of any lack in the *essentials* of an engineer's training, the fact has not come to our knowledge. The exhibit in the east gallery is designed to show that the Institute is not a shop with a school attachment.

The graduate receives the *degree* of Bachelor of Science in course. This is a short way of saying that he has had a mark of at least 60 at each semi-annual examination, and has completed the course. Then we offer a degree which shall recognize professional success, to every graduate who in actual life achieves that success.

By the blending of school and handicraft we secure an equality of social regard, which is a matter of great importance. There is no chance for the scholar to discriminate against the workman, nor for the reverse. There has never been any caste feeling in the Institute.

The only electives are courses rather than studies. A boy may elect his course, but cannot elect his studies. We deem it to be wholly for the youth's advantage that men who have experience and knowledge should devise the means and method of his training, rather than these should be remanded to his own unenlightened tastes.

What the graduates have done will be seen by examining the catalogue. Some teach, some become draughtsmen; many become journeymen. The chief peril they encounter lies in the temptation to devote themselves to earning the immediate dollar rather than the remoter fortune. Many a good man has been drawn into the office as a draughtsman, from the bench, where a true conception of his own highest interest would have kept him. But all these difficulties time will remove. Success somewhere ought to be assured to graduates in the act of graduation, provided always success is sought with clean hands and a pure heart; but success in becoming superintendents and foremen cannot be insured. Many a youth errs greatly in fixing his eyes upon this as the only success, forgetting that it is not *what* a man does, but *how* he does his work, that is the *crux* of a successful life.

It seems to me that this question of technical education is of sufficiently serious import to warrant the attention given to it by this body of engineers. Educators will read and ponder your remarks and suggestions with extraordinary interest; one of them, at least, with gratitude, for it is a matter of no ordinary moment to take into one's hands an intelligent, enthusiastic youth, and to so direct as to practically control his future.

THE CHAIRMAN: We have but a little time before we must adjourn, and you will regret with me that the discussion which has become more and more interesting, will have to close so soon.

Out of a number of gentlemen who agreed to take part in our proceedings, some were absent last night, or their names would have been called; and some are still absent. Several have furnished in writing the remarks which they would have made, and I earnestly hope that those gentlemen who are obliged to leave to-day, before they get opportunity to speak, will take the same course. It is the intention of the joint committee to publish these proceedings.

Every gentleman who takes a part in this discussion will have the opportunity to revise his remarks before final publication. Those who for any reason do not speak, are requested to send to Mr. Holley, the secretary of the committee, their views written out for publication.

MR. P. H. DUDLEY, *Member of the Institute of Mining Engineers.*

MR. CHAIRMAN: Having been asked to discuss this subject with reference to railway engineering, I shall assume it to be granted that thorough technical culture and training are necessary in all departments of this branch of industry, and I shall consider the question—how and when can such education be best imparted?

Technical education implies in its fullest sense a thorough knowledge of scientific principles, and their application in practice. In the question under consideration, by what is really a misunderstanding, we are apt to consider, especially in our youthful days, the two great means of education, viz., our schools and the railways themselves, as having a repulsion instead of an attractive affinity, whereas, in fact, the practice of the railways is but the higher development of the school,—the utilization of the principles taught in the school; so that the want of harmony between them is more imaginary than real. Railway work has such a broad and comprehensive meaning that it includes in its detail more or less of all the distinctive departments of engineering; also, general science, law, finance, and not least, a knowledge of the condition of the people. It is a recognized fact, that the men who have the best understanding of all these, and the nerve and ability to execute, are the most successful railway managers. It is true that railway work is conducted under separate departments, yet the man in general control should have such a technical knowledge of the respective details that he can detect an error in administration, and give general directions for its correction. I need not tell you that this knowledge of details, and ability to execute, can only be acquired by practice; it cannot possibly be acquired from theory alone.

Our limited time of ten minutes will not permit us to go into the details of all the departments of railway work, yet from a few explanations it will be readily seen that some branches of the necessary education can be more easily imparted in the school than others—from the fact that it is within their scope to combine the theory with the practice—such as chemical analysis, testing the strength of small materials, etc. But when we come to the erection of great structures, preparing the foundations, executing engineering works, and conducting their operations, we pass away beyond the scope and ability of any school to furnish the means for such *experience* as the engineer must have. No matter how well he may be versed in theory, his faculties must be so well trained by the discipline of experience that in an emergency he can intelligently and calmly command himself and others. No matter how well the plans may be prepared for a particular foundation, something often occurs entirely unexpected, requiring quick and decisive action, which is only possible to the practical man. It is in theory very simple to make a reconnaissance for a railway, but it takes long practice and education of the eye to enable one to judge correctly of the many lines before him, retaining their features in his mind so that he can recall and pass them in review for comparison with other lines.

In construction one can acquire in theory but a small part of the knowledge which is necessary for successful practice. Timber and other material is not usually found the same as described in books, yet the best effect must be produced which the circumstances allow. The faculties must be so educated by long familiarity that one can take in the situation at once, and decide upon the fitness of his material by intuition as it were, and not be obliged to spend days to test every unimportant piece of his structure.

In public works, the frequent want of honor, on the part of persons connected therewith, increases the difficulties of construction to an alarming extent, requiring much practical experience to judge of the quality of work and to oppose the guiles and flatteries of those who are interested therein. Questions are constantly arising which were not even treated in books—the solution of which not only involves principles of construction, but those of honor and reputation, requiring the engineer to act as an impartial judge between parties. Is it to be expected that men will be put in charge of such works without experience in such matters? In railway operations especially, those who conduct them must have experience—must be so trained and disciplined in the very work they are doing that they are masters of the situation. Accidents from the elements and from human weakness, carelessness, ignorance, stubbornness, and malice are liable to occur at any moment. In view of all these constant dangers to life and property, railway managers feel their responsibility, and they do not call men to fill such places until they have been tried and tested, in the progressive school of experience and practice, and found not wanting.

Besides knowing the principles from a study of theory, it is necessary to educate the faculties, so that impressions will be distinctly seen and properly comprehended. There must be such broadness of view that superficial conclusions may be avoided. One must have such training as will correct the ordinary aberrations of unskilled sight; the touch must have its sensibilities increased, the ear must be trained—in short, all the faculties must be on the alert to properly direct engineering work. This can only be acquired by practice and experience, the length of time depending upon the ability of the subject.

It is a delusion to hold up to young men the idea that, as soon as they have completed their theoretical education, and become conversant with the principles of science, they will be called to places of responsibility. Such men are only partially educated.

On many railways they have a system of promotion, advancing men as they acquire experience to higher positions, but, in every instance, to manage departments with success, the man must have also the technical education, and if he has not acquired it in the humbler situation, the railway company will be taxed to pay for it. Take the car builder, for instance. If he does not know how to construct his cars with the least amount of friction, the companies pay for his want of knowledge in the increased cost of running his cars. If he increases the dead weight unnecessarily, the company pays for moving it, as well as the increased cost of construction. Take the master mechanic. If he has not brought his motive power up to its highest standard, the company pays for the want of his education in the increased cost of doing their business. If the civil engineer has not adopted the most judicious grades, and the shortest line, the company pays for his want of education in the increased cost of operating their road. It does not matter what officer it may be, if he does not recognize and follow out correct principles, some one must pay for his want of education. Principles of science are not invented; they exist in conformity with unchangeable laws, and are not subject to the dictation of any man, however high in authority. It is true that art has preceded science in all cases, but its development is in the ratio of the knowledge and application of scientific principles.

It is less than half a century since our vast railway system was inaugurated, and from the constant improvements which are being daily made, no one presumes that it is near perfection, therefore we must not expect that the science has been fully elucidated, and its principles pointed out. The art was practiced long before any systematic inquiry was made into principles. To determine these requires the collection of facts pertaining to all the operations of conducting railways, and arranging and tabulating the results. Now, while a vast amount of information can be obtained, much that is needed is not even kept by railway companies, and what is kept is merely approximate, from which correct conclusions cannot be drawn. If railway companies, who have the facilities for collecting correct data, so as to profit in the greatest degree by their own experience, fail to do so, is it just to expect that our schools, without facilities, can collect this data, and formulate the results for railway men? One great reason why theories of railway operations oftentimes prove so futile, is, that the conclusions were formed simply from a mass of error furnished by the railway men. Are the school men responsible for this? To educate men who are to control railway affairs, it is clearly the province of the schools to furnish them such a training that they may know how to think upon a broad, theoretical basis. The schools must develop the powers of thought so that principles can be recognized, and a still higher discipline will show how to discover principles. It is clearly the province of the railways to teach adaptation and utilization.

The same considerations are equally true in regard to the subordinate men who form a part, as it were, of our railways. It is only within a few years that any theoretical education was considered necessary for the ordinary railway man, but now that it is, employés, encouraged to some extent, have formed organizations to assist themselves in gaining the theoretical knowledge. Railway companies to some extent give premiums for the most efficient services, and find that such investments pay well.

We must not underrate the great and noble work our schools have accomplished in liberalizing thought and advancing civilization, and in thus indirectly promoting the development of engineering. It is only important for teachers and students to remember that the graduate is but half educated, and it is equally important for all the workers, however high or low their position, to remember that but half the elements of professional progress can be derived from unaided practice.

PROF. J. B. DAVIS, *Junior of the American Society of Civil Engineers.*

MR. CHAIRMAN: It is with great diffidence that I rise here, a mere boy, to speak to this audience. Those who have preceded me have anticipated already what I had intended to say. I must beg leave to call your attention more to my manner of looking into the subject, than to anything new I may present that bears upon it.

It seems to me that we labor under difficulties, in the matter of technical education, precisely similar to those that are of a fundamental character in education in the schools in general. We are in danger of doing, indeed, we have been doing, the same things that are and have ever been done by educators since schools were. There has been a world-wide attempt for centuries, to substitute the training of the intellect for the complete training of the whole individual. Let us for a moment consider how many there are of the millions that throng the earth, that, being called upon in the daily affairs of life to take action upon any matter, will act from previous intellectual training. Certain it is that their training in this respect does assist many in forming their opinions or in conducting their operations, but this is only an item in the great mass of activity. Consider the myriads of *little* deeds, if I may call them so, that go to fill up the sum of the work of each one of us, and to really make our lives what they are. Does intellectual training furnish the mainspring of effort here? Certainly not. We do what seems to us to be best at the moment, according to the way in which we are affected emotionally, and not intellectually. I think we may well ask this broad question concerning all our usual schooling: Is the most necessary part of our education attended to? If we do not usually act from the motives furnished by our intellectual training, are we wise to devote our whole energies to the training of this kind, to the almost total neglect of that part of ourselves from which the real motives of action arise? There will be much said during this year of grand celebrations and general congratulations upon the completeness, worth, and high character of our well-nigh universal school system. I am one of those, too, who are ready to regard the results attained with great general satisfaction. There has been an immense and valuable work done, but it seems to me to be a matter of the greatest consequence and significance that so much has remained undone, even almost untouched.

Now, as regards technical education, for I cannot dwell longer on the interesting considerations hinted at above, we are doing in a measure a similar misapplied and defective work. Each one here knows of many instances in which young men ostensibly prepared for some department of practical life in the schools have failed partly or entirely owing to improper training. Indeed, the general expression of opinion here given has been one of distrust of a simply school-educated young engineer. Now what is the trouble, and wherein do we find the difficulty? The trouble is in seeking to make one kind of training answer for another, or in exclusively training one set of faculties when another should be trained. The difficulty lies in copying too closely the theories and methods of the schools in general. It is a well-established fact, that a technically educated person must possess judgment in the line of his work. This has been repeated and reiterated here. Who supposes that he will possess that judgment if he never has a chance to exercise it? If I stand over my boys and tell them here, do this, or here, do that, they will never be *judges* of their work. They must be set by themselves to execute plans alone. Another attribute of a good engineer, if I may refer to him as one of the technical class, and one which I have not heard as much as I wish about, is that of resources. He must be a man of resources, of invention, of ready expedients. Can a man hardly be an engineer at all without this invention? This thing has been incidentally referred to in many of the communications, but I wish to emphasize it. The engineer must be a man ready to act whenever he is called upon within his field, in all emergencies, and on all occasions. The young men are to fit themselves for emergencies, for ready, reliable action on short notice, or under difficult conditions. Who supposes for a moment that a man will possess resources without ever having faced a difficulty? These two illustrations indicate in a way, the nature of my theory of the training a technical man should have, which is this: Consider well what characteristics he should possess, and then seek to develop them in him if he has the basis. If not, tell him so, and not prepare him for the great disappointment of losing his years of study and work in a great failure. As to how this result is to be reached, I will say this: At the University of Michigan we have met the difficulties alluded to, that is, a disposition to follow old methods and theories of intellectual training. It was supposed that an engineer must be well developed intellectually in certain directions. Therefore certain studies of the older courses were retained and advanced, and others added to give him what was deemed a desirable power in these directions. Much was done of a *practical* kind, so called. For years they have been feeling their way along at this institution, seeking by thought and experiment what was best. And in my own experience I will say that I have been satisfied with the result of my teaching about in proportion as I was able to realize the conditions of actual practice in my instruction. Personally, I think the same is true of the whole work of the engineering department, except I will add that these conditions were supplemented by good class-room instruction in theory. And as to the manner of combining theory and practice, I have but this to say, although much has been said upon this head; I think the more intimately they are combined daily, the better.

As to mathematics, which have been mentioned, we think, at this school, that there is such a thing as enough mathematics for civil engineering students. The conclusion has been reached that there is a point where the course in these studies should stop. Perhaps our conditions are somewhat different from those of the gentlemen who have preceded me. A certain amount of mathematics well mastered is what we desire. A limit has been reached with us for our undergraduates.

I would like to say a few words about the study of drawing. Instruction in drawing should not be postponed, as is frequently the case, till the student enters college. It should not begin in the high-school, nor even in the grammar-school, for this reason: it is, crudely, to be sure, a more natural mode of expression than writing or even talking, yet this method is left to be taught long after these others have been acquired. But I shall be told that the others are absolutely necessary in the common affairs of life, and must be learned. True, yet how many educators have considered the relations and mutual helps which exist amongst the different methods? The absolute "bread and butter" importance of drawing is just being felt and realized. It is a matter of common occurrence to see two persons conversing, and what one cannot make clear to the other by words he will seek to mark out, it may be with a stick in the dirt. How readily do we all seek this method of expression. Much more can be said about its natural relation to us. It certainly seems that the child should begin drawing soon after learning to read easy words. The Kindergarteners begin it really before this.

As to the young men having command of their knowledge, which the gentleman from Worcester referred to. When the engineering students reach their junior year with us, it happens that I teach some little things then calculated to open their eyes to the fact that there is quite a difference between learning a thing well enough to make a recitation upon it, and mastering it so they can use it. Sometimes it takes the third trial to get a good recitation at this point in their course, and they generally have quite an experience during those three days. The lessons are long and the parts so similar that memorizing is a hopeless task. They must *master* the work in order to recite. The test applied at this institution to find out whether a man knows a thing is, can he do it? So far as possible this is everywhere carried out. It is not enough that he can converse about it. Can he do it?

One word upon the item of integrity, and I am done. The only thing I will say is to relate a short anecdote. I once had an assistant whilst continuing some railroad work from previous surveys by another man. This assistant had been upon the work before I was, and had worked upon the preliminary lines. He was the leveller. I was locating over his preliminaries. As we worked along, a number of errors were detected in our levels. These kept accumulating. Upon investigation I found that this assistant, who was still running the level, had made an error of twenty feet upon his previous work. This he had nicely distributed through his notes, so that it did not at once appear. This was far away in the woods of Northern Michigan, where it was impossible to obtain a man for this kind of work. The way we prevented him from doing more mischief was to set the rodman to watch him. Each kept a separate set of complete notes. They were not allowed to show their notes to each other till the day's work was done. Then they were obliged to go through them all in the evening, and compare them carefully. In this way we fixed our leveller so as to keep him honest till we arrived where we could get rid of him. I think I tried hard enough to teach that man the error of his ways. Some years after, I received a letter, to my entire surprise, from my delinquent assistant. He was then a resident engineer in charge of 600 miles of line in Minnesota, and he said, "I owe to you, Mr. Davis, my position, because you taught me by your severe treatment," in the case referred to, "the value of integrity." He further said, "After I left you I went West, and went into an engineer's office as a draughtsman at a dollar and a half a day, and in these few years I have come to be in charge of 600 miles of line." Let the story carry its own moral.

Mr. Frederick J. Slade, *Member of the American Institute of Mining Engineers.*

Mr. Chairman: If it were certain that all the young men who enter our technical schools had the natural qualifications necessary to be engineers, then the problem, of what the course in such schools should consist, would be materially simplified. But when we remember that it is impossible at the early age at which young men enter on such a course, to determine accurately the natural bent of their minds, the necessity of first imparting a liberal general education is apparent, so that those unfit for engineering pursuits may have other fields opened to their view, and may be drawn away from a profession for which they have no fitness. I therefore agree with the remark that has already been made, that if it were necessary to choose between a strictly technical education and a more general course, the latter would be the more desirable.

I believe further, that not only is it desirable on account of the various types of mind to be found in a body of undeveloped young men, that education should be general rather than special, in order that none may be graduated without having received a training which shall be of service in his particular case, but that even were the classes composed of none but those qualified to become engineers, it would be much better that the instruction should be confined mainly to the theoretical part of the profession, leaving the practical details to be learned afterward, in that school of actual practice from which the engineer never graduates. I know that this would require some of our schools to give up some of the very things that they most pride themselves upon, yet I believe the effect in the end would be good.

I think it has very generally come within our experience, that those engineers who have received that very elaborate education in foreign schools, of which so much has been said, do not make the most rapid progress in the practice of their profession. The effect of their study seems to have been to give them a disproportionate confidence in the sufficiency of the formulæ and text-books of the school, to solve every problem that arises in practice, and the faculty of judgment so necessary as a check upon theoretical deductions becomes dwarfed by disuse. It sometimes seems even to be the case, that those who have received this training refuse to admit the necessity of correcting their theory by facts, and hence shut themselves out from the greatest of all schools.

It has been remarked with force by Prof. Thompson, that the acquisitive powers act with greatest vigor before the age of twenty; and he argues that on this account both theoretical and practical education should be crowded into this period. While it is no doubt true that the mind is at this age better able to receive scientific education, and to be moulded by it to correct methods of thought, it may be doubted whether the ability to weigh the value of practical expedients is as much a characteristic of a young as of a more experienced mind. It would, therefore, appear to be wiser that the period of life in which the mind is best adapted to receive that scientific training which is to shape all its future action, should be devoted to the study of science, rather than wasted in the vain attempt to acquire an incomplete and delusive acquaintance with practice. One very important result of this would be, that the young graduate could not by any possibility imagine himself an engineer, and would thus be ready to commence at the bottom, and on a true foundation lay up in intelligence and order, that experience which is the capital of the engineer.

It may be asked, why should not the teaching of practice *with* theory have the same effect in establishing a proper balance as when acquired subsequently? To which it may perhaps be answered, that the conditions under which the practical matters are presented are in the two cases widely different. In the school they are presented as problems solved, in actual practice as problems to be solved; and the fact of the solution being at hand in the former case gives a false idea of complete mastery of the profession, the precise reverse of that modesty which is forced upon one, in the latter case, by the uncertainty whether he shall be able at all to reach a solution.

As to the proposition that the student should in the midst of his theoretical course, take up the study of practice as presented in the workshop, I think it is a question whether the time so spent would not be spent wastefully. It is indeed highly desirable that the young man while pursuing scientific studies, should be sufficiently familiar with at least the surface facts of practice to give life and meaning to the abstract truths which he is studying. But this acquaintance he can acquire in those afternoon visits which every young man, having a taste for engineering, is sure to make to such works as are within his reach (and in these days there are a multitude in all places), while he could acquire but little more by a constant attendance in the shop, *unless in actual employ and with a weight of responsibility resting upon him.* It is only when the sweat comes out over a man in those emergencies, when he knows that something must be done, and done quickly, that he begins to lay up valuable experience.

Now, it would be impracticable to provide such employment for students; and I think it may also be said that in nine cases out of ten, where it could be provided, the young man would never return to school, because the ties that would bind him to his actual work would be too strong to be severed.

In a word, then, let the schools give a liberal and scientific education; let the student give concrete form to abstract principles as he may from visits to such works as are within reach, and by the reading of current engineering literature, and let the acquisition of practical knowledge begin and go on without interruption after the school course has ended.

MR. ASHBEL WELCH, *Member of the American Society of Civil Engineers.*

MR. CHAIRMAN: I move that the Joint Committee be requested to take into consideration how to put a stop to the practice of the technical schools giving the young men who have just graduated the title of civil engineers.

In the first part of my remarks last night I had this paragraph: "The young man that graduates from a technical school is no more a civil engineer than a young man who has never been at sea, but who has studied navigation at school, is a sailor." The same thing has been better expressed by Mr. Coxe, that a graduate is excellent raw material to make an engineer of, but he is not an engineer. Now you confer this title when these young men graduate, and what is the result? Why, they think that they are engineers, and the consequence is that in nine cases out of ten they will never learn to be engineers.

Another thing is, that to those that really are entitled to that name, the title of civil engineer itself becomes meaningless, from the fact that one of these boys that are called civil engineers is put upon a par with the man who has been in the profession thirty or forty years, and that title is of no use to him. Whether it is practicable to stop it I do not know, but the sooner it is stopped the easier it is stopped. When we have gone on for four or five years in doing this thing, it is very difficult to stop it, more so than if it had only gone on four or five months.

A MEMBER OF THE AMERICAN SOCIETY OF CIVIL ENGINEERS: I would second that motion. I come from an institution that does that thing, and I wish it could be abolished.

THE CHAIRMAN: The chair is in doubt as to the legitimacy of the motion. At present the chair decides that it is not in order. This meeting is a meeting held by the two societies jointly for a special purpose. In order to arrange for it, a joint committee has been appointed. The chair is of opinion that this meeting as such, cannot perpetuate that committee nor direct it to consider any subject. Each of the societies must act for itself on such a question, and, if a joint committee is advisable, then it must be created by the separate action of the two societies, each according to its own rules. But this decision of the chair need not detract from the moral effect of the excellent suggestion made in several speeches and by Mr. Welch, and by the seconder of his motion, in which I heartily concur. I should be glad to see the granting of degrees restricted in accordance with the suggestion. Meanwhile, unless some gentleman desires to appeal from the decision of the chair—a step which will be welcome, as relieving the chair from responsibility in the matter—this motion is ruled out of order.

Gentlemen, in concluding this most interesting and timely debate, permit me to call your attention to what may be called its net result. Several things seem to have been settled by it, besides the innumerable things which have been thrown out to stimulate further inquiry and debate.

We have settled upon the necessity of a wide education, if we wish to raise a high standard. We have settled upon the necessity of practical training, somewhere in the outset of the work of the young engineer. How shall we get both of these things? We have got to learn more than our fathers, and yet not live longer. As Prof. Thompson has forcibly shown, the period of study is limited by nature. The necessary time and efficiency is not going to be gained by taking the old curriculum and shifting from one thing to another, by giving up the old-fashioned studies and putting in a little of the new-fashioned studies. Change in methods, not merely subjects of instruction, will help us much. I am satisfied that there is an immense amount of useless labor put upon boys, under the name of mental discipline. Take the study of Latin and Greek, for instance. Two-thirds of the time employed in getting the lesson every day is spent in the turning over of the leaves of a dictionary—mere mechanical exercise. Yet it is thought dishonorable to have a "pony," or translation, which is nothing in the world but a dictionary so arranged that you don't have to work your elbow to find words, that is all. The method of teaching modern languages is much better, and it ought to be applied to ancient languages. We must teach them quicker or give them up. Similar reform can be carried even into exact science. Take, for instance, Professor Thompson's remarks in respect to algebra, which are exactly in point.

Another matter of great importance has been suggested and universally admitted, namely, the difficulty of knowing beforehand whether young men are fitted for what they are going into. Nothing is more difficult than this. Life is full of men who have mistaken their vocations. One of the best things about a preliminary course of practice for engineers would be, I think, that men naturally unsuited for engineers would be sifted out before it became too late for them to change their plans and preparations for life.

Now what is the testimony of this debate as to the actual result of the experiment of preliminary practical training, so far as it has anywhere been tried?

When men like Professor Thurston and Professor Thompson, who have distinct plans, and are engaged in really scientific experimenting in the matter, come forward with their results, they are entitled to great respect. They tell us just what we want to know. Professor Thompson's testimony as to the relative progress of those who came to the Worcester Free Institute after an apprenticeship in ordinary shops, and those who came without any apprenticeship, and took the course in the workshop of the Institute, is, I venture to say, the hardest blow that Mr. Holley's plan has yet received from any objector. The opinion of other instructors on the same point would be valuable. But it is only fair to say that Mr. Holley looks to something more than an ordinary apprenticeship; and so the comparison drawn from experience at Worcester is not wholly fair to him. It should also be remembered that the establishment of such training-shops as those at Worcester, and Hoboken, and Ithaca, and some other places, requires money endowments of no mean extent, while Mr. Holley's scheme can be tried, where it is applicable at all, in commercial establishments already existing.

The meeting was then adjourned *sine die.*

COMMUNICATIONS.

THE following papers, intended to form a part of this discussion, were received before or during the meetings. They are appended in alphabetical order.

PROF. RICHARD AKERMAN, *of Sweden.*

In Sweden all education is free, all the teachers being paid by the government. Some of the technical schools, however, were originally founded by private means, in consequence of which their teachers get their pay partly from the government and partly from the income of private donations. According to the different degrees of knowledge required for admission, the technological schools in Sweden are divided into three kinds.

For admission to the schools of art, the lowest of the three, no more knowledge is wanted than what is acquired in the ordinary public schools. The most prominent of the above named, the School of Art in Stockholm, has about 1700 male and 800 female pupils.

Between the schools of art and the Technological Institute in Stockholm, which is the highest technical school of Sweden, there is an intermediate kind of technical schools, which have more moderate terms of admission than those of the Technological Institute, but much higher than those of the schools of art.

For admission to the Technological Institute, in which the School of Mines is incorporated, the amount of knowledge required is as follows: two grades of algebra, trigonometry, stereometry, geometry, and the elements of chemistry and physics. In general knowledge, the absolutely necessary requirements are very insignificant, as it is not important to have more than a passable knowledge of the history of Sweden, and except the Swedish, of one other living language, usually the German. Nowadays, however, by far the greatest part of the pupils have passed the high schools, and therefore have a thoroughly good general knowledge. The number of pupils at the Technological Institute is about 270. The course has been, and is still, three years, but last winter it was decided to make it four years, this to commence from the beginning of next year.

The last year of the course especially, has a practical tendency, the teaching being altogether in applying, and the reason for the above-named decision, to make the term hereafter four years long, is to be sought for in the circumstance that two years have been found insufficient for acquiring a thorough knowledge of mathematics and the natural sciences, which form the foundation of the technical education, and also proper skilfulness in drawing and constructions.

When, as mentioned before, the teaching during the last year is exclusively in applied sciences, the natural consequence is, that it must be essentially diverse, depending upon the branch of engineering the youths are going to embrace, viz., civil, mechanical, architectural, mining and metallurgical, or chemistry. As, however, different degrees of knowledge of the natural sciences are required for different branches of engineering, the teaching, with due regard to this, is alike for all the pupils, only during the first year, and at the beginning of the second year, they must make up their minds to what branch they intend to devote themselves.

The School of Mines is exclusively a school of applied science for those who desire to make mining or metallurgy their special study. The teaching in this school, therefore, corresponds with the last year in the Technological Institute, and the requirement for admission consists of that particular knowledge acquired by those who at the Technological Institute have devoted themselves to mining or metallurgical engineering.

By far the greatest number of ordinary pupils in the School of Mines have therefore finished their preliminary studies at the Technological Institute, but those also who have passed through lower technical schools, and therefore have not taken such extensive preliminary studies, are allowed to enjoy the same teaching as extra pupils, though they are not allowed to pass any examination, and do not get any certificates.

Besides the preliminary theoretical studies, it is stipulated as a condition for admission to the School of Mines, that the candidate for at least two months shall have lived at some mines and iron-works, and during this time shall have tried to the best of his ability to understand the different processes and work there going on. For this purpose the time from the end of the term at the Technological Institute (12th of June), and the beginning at the School of Mines (8th of September), is used; but nearly every year there are several pupils of the School of Mines, who, after they have finished their terms at the Technological Institute, have taken during several following years situations at mines and iron-works, and not until they have acquired practical experience in order to become more conversant with the theories of the different processes, do they go to the School of Mines.

From the 8th of September to the 1st of May, the teaching in mining and metallurgy, chiefly that of iron, is going on without interruption, in Stockholm, and even the first practical knowledge in mine-surveying, is obtained in the rooms of the school itself; but on the 1st of May the pupils start travelling, and at first pass six weeks at some iron-works in company with and under the superintendence of their teachers. There they have to blow-in a blast-furnace only sixteen feet high, belonging to the school, without any help whatever from professional or other people. Then, they are called upon to produce different kinds of pig-iron suitable for different purposes, and they have to calculate themselves the proper mixtures to obtain this result.

Besides this short blast, which they manage altogether on their own account, they are obliged to watch the working and management of the large blast-furnace belonging to the works very closely, as well as to take part of the manual labor at the ore-calcining kilns and the Lancashire forges.

Having finished these practical studies in blast-furnace working and the process of making malleable iron, they at once proceed to a mine, where they, under the supervision of another teacher, practice surveying, and study in a practical way the extraction of the ore and the general working of the mine. This is continued till the beginning of July, from which time the pupils, under the guardianship of their teachers in the respective branches, visit the most prominent and notable iron-works and mines of the country. These travels go on till the end of July, when the pupils have to write down an account of them, and the impressions received during this time. The nature of these accounts are taken very much into consideration by the teachers, in the determination of the certificates. The final examinations take place at the end of October.

MR. CHARLES MACDONALD, *Member of the American Society of Civil Engineers and of the American Institute of Mining Engineers.*

A recent writer in the *New York Times*, in speaking of education, has said: "Few things have been made plainer within a few years than the falsity of our system of education as generally followed, the chief defects just now being in *method* and *discrimination*."

Judging from the tenor of the discussion of Mr. Holley's able paper, this view would seem to be very generally held by members of our own profession in regard to technical education; hence the elaborate comparisons of different methods, old and new.

There have been advocates of a purely theoretical system, by which general principles are taught without the aid of practical illustration. Then, again, it has been suggested that the methods of the workshop may with advantage be introduced into the school, and lastly, as Mr. Holley indicates, there is the possibility of reversing the last method, by bringing the school into the workshop.

It does not seem to me that the inadequate results which are too often obtained from our technical schools are due so much to the methods employed as to our own treatment of the individual with whom we have to do.

Between the ages of ten and sixteen the mind is in a condition to receive impressions only; it is, therefore, eminently proper that this period of life should be devoted to the acquirement of what is generally understood as a common school education, in which I would substitute French and German for the dead languages, which have heretofore been so persistently and indiscriminately forced upon the pupil. At the age of sixteen the body is in much greater need of attention than the mind; it may be said to be in a transition state, in which relaxation and change of scene are of the greatest importance. The boy should be taken from school at this age, and allowed a vacation of from six to twelve months, during which time he should be permitted to see as much of the world as practicable, and should be given to understand that he has now arrived at that most important period of life when he must select for himself an occupation. At the end of his vacation, if his preferences are decidedly in favor of the engineering profession, he should be allowed a year's practical contact with that branch of it with which he proposes to connect himself. If he is of a mechanical turn of mind, it would not appear to be difficult to obtain a position for him, without pay, in some manufacturing establishment, where he would be daily brought in contact with the best that is being done, and with the men and methods by which it is accomplished; he would begin to recognize the difficulties under which practical men labor, and at the same time he would see the necessity of a knowledge of practical methods, to the successful application of general principles. This experience would undoubtedly inspire him with a desire to acquire a knowledge of such general principles, and would furnish the *incentive to study*, which is, after all, the one thing needful; without this the best methods may too often be employed without imparting knowledge, while, on the other hand, if the student has been enabled to see the end from the beginning, he will absorb much that it is important for him to know, even under the most defective system. Great care should be taken in limiting the time spent in preliminary contact with practical life, lest the habit of study be impaired or the student shall have been impressed with a desire to accomplish results without a proper appreciation of the importance of scientific methods. I have mentioned a year as generally sufficient for this introductory work, at the end of which time the student, at the age of eighteen, will be in proper form to enter upon a four years' technical course in a purely theoretical school. This should bring him to the full stature of manhood, well equipped and eager to encounter the difficult problems of an active professional life, and with a store of knowledge at his command, which will enable him to render a good account of the abilities with which he has been endowed.

To conclude briefly, I would adopt the views expressed by Mr. Holley, in so far as they indicate the necessity of a short, practical experience, before technical training begins, but I believe this can best be accomplished by placing the student in a commercially successful manufacturing establishment, or upon some scientifically conducted public work, without pay, and under the direction of the managers of the work solely. Contact with other students, or with expert school-men at this stage, would divert the mind from the main object, which is to acquire a correct idea of what is being accomplished in practical life, and would not tend to develop the habit of self-reliance, or the improvement of the mind in the direction of its greatest capability.

MR. ISAAC NEWTON, *Member of the Institute of Mining Engineers.*

Sound acquirements in such thoroughly practical professions as those of civil, mechanical, or mining engineering, it seems to me, can only be attained by actual practice; but, before the pupil can take part in real operations, or can even see them with any profit, it is essential that he should be thoroughly grounded in the ordinary academic studies; but he should also have obtained such knowledge of the profession he is preparing himself for, as can be acquired by instruction in a technical school. A pupil at the beginning of his career, after leaving the technical school, finds out, as soon as he begins to face the realities of his profession, that, instead of having completed his professional education, he had only begun it. From what little I have seen of the course of technical students, I am disposed to think that the first two years or thereabouts, after graduation, will give the young man more accurate knowledge regarding his fitness for his work than can be got from the school itself.

It seems to me that it would be a wise plan for our polytechnic schools to divide their degrees into two steps or stages; the first to be bestowed on the student on graduation, and the second after an examination, to be held after, say, the first two years of his course. During this period the candidate should keep a record of his technical reading, with his observations thereon, of the engineering operations he has witnessed, and of the work he has been engaged upon. The character of these records, and an examination based thereon, should be the ground of the award of his final or full degree. I feel convinced that if this plan, or one similar to it, could be carried out, a few years would witness a marked improvement in the rising generation of our engineers, civil, mechanical, and mining.

There is another matter, it strikes me, which ought to receive more attention than it does in the polytechnic schools. I mean the purely business training that is essential to financial success in all the professions under discussion; a knowledge of bookkeeping, and the methods of making up accounts, will give the engineer who has this knowledge a vast advantage over one who is deficient in this particular. I have on more than one occasion seen young men about iron establishments and engineering workshops, whose sole technical knowledge was acquired by their own unsupervised reading, and by watching what was going on about the works, promoted over the heads of those who had graduated from technical schools. The young graduate, full of enthusiasm for his profession, is too often prone to look with a sort of contempt on such acquirements, but he may rest assured that a familiarity with business methods is a vital necessity for his permanent success.

Let me call attention to one more point. All will admit the vast importance of accurate knowledge respecting the cost of production, and the comparative cost of various methods and processes; all will at the same time admit the difficulties of finding this out by watching the operations of various works. Pupils cannot be too strongly impressed with the importance of such knowledge, and they should be taught to carefully utilize every opportunity to get it.

MR. CHARLES B. RICHARDS, *Member of the American Society of Civil Engineers.*

What follows refers only to mechanical engineering, particularly to machine construction, in relation to which the subject under discussion is of the greatest importance.

The system in large American shops (particularly in New England) of the subdivision of labor, and of employing one man on one kind of work only, is so prevalent, that good, practical, general mechanics are not being produced. The system of employing apprentices who shall learn general work in the machine shop is in disfavor, first, with the employer, because to carry it out fully does not correspond with the system of subdivision; second, with the employé, because skill in one kind of work is soon acquired, and this done, he finds the tendency of the subdivision system is to keep him engaged on the one kind of work he has become skilful in.

The places of foremen and directing mechanics who, from the general character of their training and practice under the old system, became fitted for their positions, cannot be filled from the ranks, in shops where the system of labor subdivision prevails. Good planer hands, or lathe workmen, or fitters, etc., are being trained from men who enter the shops at low wages, and devote an indefinite time to practice in a single process, in which their labor soon becomes remunerative, and to which it is therefore devoted exclusively. Hence, a general knowledge of processes is not acquired. Our directing mechanics and engineers, who in former years, and almost up to the present time, were raised from the bench, must hereafter be trained in the schools, and it is to be confessed that their education is a perplexing problem.

President Holley has, in my opinion, touched the keynote, in assuming that practical instruction should precede that of the scientific school, for an interest in the sciences, and facility in acquiring them, are developed as the application of the sciences to the arts becomes apparent, and are increased in proportion as these relations are more evident.

What the nature of the preliminary practical instruction should be, is the important question. It should be of a character to insure the greatest saving of time to the pupil in attaining the object of his aim. It should not extend over so much time that the pupil will be deterred from attempting the scientific school course. It should be strictly preparatory to the theoretical course.

I do not think this preliminary study should consist of, or, indeed, should necessarily comprise practical work with tools in the shop, for so many arts are here brought together, that almost the first question which arises is, which one of them shall be practiced? Shall it (for example) be that of the pattern-maker, or of the moulder, or of the machinist? It is evident that a general knowledge of all is essential to the engineer, and a certain amount of skill in all, perhaps not in one more than in another, would be desirable, but can the time be afforded for this? It is a question whether, in a course of preliminary study, opportunities for general observation, carefully directed, are not more useful than those for putting one's own hand to the plough.

I believe the best training, preparatory to the theoretical course, can be obtained by about two years of hard work in the general duties of the office of a directing engineer of works, who should direct the instruction, and should of course be a cultivated practitioner, who will look after the interests of his pupils. In such a position the student can be given the best of opportunities to study, at the drawing-table, the details of machines and of processes, the construction and carrying out of which he ought to have opportunities of attending to carefully in the shop. The discipline of the office will be most useful to him. Drawing, one of the arts he is to practice, will be acquired in a way which will serve to give a clear and practical aspect to subjects whose mathematical treatment he will encounter in his subsequent theoretical studies. The time necessary for acquiring some skill with the pencil and pen will be saved from the school course, and will also be well improved in the study of approved examples of modern construction, rather than the models of obsolete practice which are found in many school cabinets.

I believe it to be hardly practicable to organize a system of very profitable preliminary instruction, in which the practical working in the shop with tools is a feature, for even under a competent instructor, most boys would need to practice elementary processes for an unprofitably long time, in order to warrant their being permitted to attempt advanced work, when the risk of a costly failure from lack of skill is considered. It is probable that many manufacturers of machinery would regard the kind of training in shop manipulation which would be the most profitable to the student, with much the same kind of feeling that the enlightened patient would regard a course of surgical practice by prospective medical students.

It is probable that a plan could be systematized for having pupils taken into the offices of engineers whose qualifications and positions are favorable for the purpose. One argument in favor of this plan is, that if, after the school course is completed, the pupil should find time for a course of practice in manipulations, he will find himself much better fitted to enter upon this with profit to himself and to the shop where he works, than he would have done when younger, and before his studies in the office had taught him where to direct his attention and efforts; and it is my opinion that, if he cannot afford the time for this last course, he will be better off without it than he would have been if this had been substituted for the preliminary office course.

PROF. R. H. RICHARDS, *Member of the American Institute of Mining Engineers.*

The two questions considered. I. Should a course of study in works precede, accompany, or follow that in the scientific school? II. Is it practicable to organize practical schools in engineering works?

The office of instruction seems to be to train the individual to observe. Observers note only what are to them the salient points of a subject. The individual must therefore be given sufficient familiarity with a subject to know what are the salient points to experts, in order that he may take note of them.

Practice before theory.—A carpenter, of twenty-four years of age, who has done nothing but mark with a square, cut off, and drive nails for eight years, has a mind used to a very few simple operations, and entirely unfit for gaining the greatest profit from the scientific school. He has a mind squared to a certain base-line, and one which can be taught only by a method which starts and builds up with reference to that base-line. Practice at the first undoubtedly brightens the mind, but when the novelty has entirely worn off, and nothing but routine work is done, the mind becomes dimmed more and more to other things and less receptive for the training of scientific schools. Men can be seen everywhere who, when once they have got into a narrow routine of work—and it requires only a short time to bring them there—are callous to all wider and larger views of their subject, and to whom the scientific-school training will have no interest or appeal, unless brought down to and built upon their own little rules of things. The order in which information may be brought to practical usefulness seems to be, 1st, observe; 2d, record intelligently; 3d, assimilate, digest, classify, compare; 4th, promulgate the results or bring them to usefulness. The routine man cannot observe nor record.

Theory before practice.—The school-man, if properly taught, has learned the principles of observation and classification, but he lacks the practice, or, in other words, he does not know what the salient points are which the expert should know, and his observation will therefore be excessively one sided, the cause of most of the so-called difference between theory and practice.

The school-man tries to "square nature with the book," and fails because he has never had sufficient opportunity to test himself. He needs to learn that his own judgment is one of the elements to be observed, and that it must be placed in the balance along with the other facts obtained. This can only be done by testing himself, and trying himself side by side with people and things. The graduate of a scientific school, unless special pains has been taken to put him upon original work or upon work in which he must *mind the helm* himself, will have to start at the very beginning, in order to make himself a thoroughly practical leader. The school-man when he finally gets to work, will find that he does not know many things which the books would have taught him if he had been awake to their meaning. This surely is not the most healthful order for gaining independence of thought.

An alternative: An educated mind, serving in engineering work, and then in the scientific school.—This arrangement will undoubtedly give the best training of all. The mind is stimulated, rationalized, and widened by a liberal education, by a free contact with men of equal age, both superior and inferior in intellect. With such a preparation, the mere routine of engineering work will not absorb and obscure the mind; it will simply be a groundwork upon which the causes and effects of things can be seen to act. The mind after such a practical training is in a thoroughly receptive condition for what the scientific school may have in store for it; the views recorded in books are brought side by side with one's own, and intelligently compared with them. But while this order of things seems better fitted than any other, it has one difficulty, and that is, the expense and time required for it are more than can generally be afforded by our young men. The question at once comes before us then, should education be planned for the few or the many?

Synchronous arrangement of practical and theoretical work.—This seems to offer the most practical solution for the education of the majority. In this arrangement the mind is constantly stimulated in its theoretical study by the practical application of what is learned. This brings the mind and soul together, strengthened by each other for the one purpose of gaining independence. There may be, however, a very considerable variation in the arrangement of a synchronous course. (*a.*) The practical side may be emphasized at the start, and the theoretical at the end of the course. (*b.*) The two may be evenly distributed. (*c.*) The theoretical side may be dwelt upon more fully at first, the practical side towards the end. My own view of the matter favors this last arrangement. The student at first should be given enough practice to stimulate his mind and explain the uses of things; but it is at the end of his course, when he is about to step away from his alma mater, that he needs to know himself, and to have proved himself side by side with people and things; it is toward the latter part of the course, therefore, that the practical part should be emphasized.

Thus far I have dwelt upon the first question, *i. e.*, that of sequence. The second question seems to me to be poised in this way: Are we to depend upon practical scientific schools, or upon educational engineering works?

In Europe the schools, mines, and works are more commonly under one control. Students can be more readily accommodated with practical advantages. In this country, however, the schools and works are carried on by different people; besides, the works are under people who believe that they possess facts which are unknown to any one else. These facts they regard in a certain sense as their capital. I fear that the time will never come in the United States, when large corporations can be made to believe that it is for their own interests to educate men who will soon be able to take charge of other similar works.

The introduction of practical work into the mining course, by the use of smelting and ore-dressing laboratories, at the Massachusetts Institute of Technology, has proved of the greatest benefit to the students. The work is upon a scale varying from one hundred pounds of ore to two or three tons, and gives a great deal of scope for testing the economy and efficiency both of the processes and of the individuals. As their laboratories have been developed from year to year, the students have steadily bettered their attainments. The work in these laboratories taken together with annual visits to mining regions by the classes, and with careful studies of the mines and works, have thus far given results which are most satisfactory and encouraging.

MR. J. DUTTON STEELE, *Member of the American Society of Civil Engineers.*

The deficiency in our system of educating young men for the pursuits of an engineer are well known to those who have been largely engaged in directing our public improvements; it is the development of the theoretical to the sacrifice of the practical. To some extent this is the case in all professions, but in that of the engineer the lines dividing the two are strongly marked, and the latter attains an importance which equals if it does not surpass the former.

The first duties of a young man on leaving college are entirely practical, and for such he has little or no preparation, and he soon discovers that he has to enter upon new fields of investigation in which he has to work haphazard and without the aid of text-books and teachers, where slow progress is made in the absence of strong mechanical proclivity, or an acute sense of observation. It is not therefore surprising that many, after spending the best part of their life for study, discover that they have mistaken their calling and give up the pursuit, whilst others, feeling that there is a future before them in the vast fields of practical science, manfully struggle with the obstructions they so unexpectedly find in their way, and after spending double the time in the school of practice, that they have already done in that of theory, find at length that they have become something more than "hewers of wood and drawers of water" to those who are their inferiors in technical education.

Practical knowledge is something more than the education of the hand or the eye of the student to skilful manipulations, it is also a knowledge of the nature of things as they exist, of the imperfections rather than the perfections of the objects with which he has to deal, and although his education is an aid to him in his after pursuits, much that he has learned has to be stored away for a remote future, or is lost in the pressing needs of the present.

For a clearer view of this subject let us assume that theoretical education is a knowledge of the forces of nature, and that practical education is a knowledge of the repellent powers in nature to the application of these forces, and we will see that the latter includes a larger field of study than the former, and one that is more difficult to accomplish, as it has not, thus far, been the subject of classification or system, and is dependent upon the opportunities and powers of original observation. Bessemer's invention for producing steel rails was so simple that any novice in chemistry could understand it, yet he had to study the repellent forces to his process which existed in the iron ores of England for seven years before he could produce satisfactory results. In the early history of our public improvements, it was not unusual to find whole series of structures failing from imperfect foundations, yet the schools had done their work well, the plans were good, but those who had charge of carrying them out had, probably, never seen a bridge founded, and were incapable of forming a correct judgment as to the requirements of the cases as they were presented to their minds in endless variety. We have had large amounts of money expended in our coal mining districts in attempts to extinguish fires in coal-mines by anti-combustible gas, overlooking the practical fact that the burning mass was generating a repellent gas in larger volume and of greater intensity than could be produced by artificial means, and which would prevent the contact of the extinguishing material with the fire. The locomotive is the offspring of practice, and it was not until it had obtained nearly its present perfection (but not its present power), that scientists resolved its several parts into due proportions, and led the way to its enlargement and greater efficiency.

Far be it from my intention to undervalue the pursuits of science; it is the only true foundation upon which to erect a structure of professional ability, but it is not *the* structure, and unless the builder learns how to put his materials together, the structure will not stand. Nor can the schools be looked to for the production of such minds as Telford, Stephenson or Bessemer; they are special creations of Providence for the advancement of mankind in industrial pursuits; they overreach the schools and conduct their investigations in new fields where they are guided largely by the lights of practice. But we should expect the schools to put forth young men capable of making themselves practically useful in the pursuits they have elected for their life-work, and to shield capital, on the one hand, from the blunders of uneducated pretenders, and on the other, from that class of scientific devotees who whittle down their theories to so fine a point that they will not stand.

This can only be done by uniting the theoretical and practical in the schools of technical education, by giving them equal prominence in the system, and by letting them go hand in hand as far as possible in the progressive stages of mental and physical development. Other systems must result in an unnecessary waste of time and money, and in disappointment to the subject.

It is however more easy to point out the defect than to discover the means of remedying it. The medical student has his dissecting-room and hospital, the chemical student his laboratory, the student of natural history his museum, all of which are complete illustrations in themselves, but the student of engineering cannot have his examples within the range of college buildings, nor can he have unquestionable control of them, distributed as they are over wide extents of country. If, however, it is admitted that the practical is of equal importance with the theoretical, an equal proportion of the school term can be appropriated to each branch of study, and a separate corps of professors assigned to each. In such an arrangement the practical teachers must be governed by circumstances during their half of the school year, but should keep the youth under their personal observation and training during the entire period. In many cases they could obtain employment for them at a nominal compensation or gratuitously, in others their time must be spent in inspection and observation. Surveying they could generally get to do, and if not, they could resurvey what has been surveyed before, and thus perfect themselves in the use of instruments and in calculating quantities; they could obtain the plans of structures, from the most simple to the most elaborate, and calculate the strains upon them, and learn that there is a practical as well as a philosophical problem involved in every operation, from the driving of a spike to the construction of a bridge, and they could enter the manufactories and mines, and gather from them a fund of useful knowledge, even if they were not allowed to assist in the operations.

If it is not considered practicable to so unite the two in a single college course, let them be pursued in alternate years, and so continued until the desired amount of education is obtained. Our present system tends too much towards binding up the minds of our young men in books, and cramps that free range of thought which is necessary to enable them to grasp the multifarious professional demands of our country, or it so depletes their physical energies that they are unfitted for the hard outdoor work of an engineer, and consigns them to the more sedentary duties of the office, in which, useful though they may be, their chances of advancement are comparatively few.

MR. A. L. TYLER, *Member of the American Institute of Mining Engineers.*

The discussion has been very interesting, and I fully agree in the necessity of a basis of general culture—now this can be laid by the time a boy is sixteen. Then, to answer the question, "What shall we do with our boys?" First, let us so prepare them that they shall get all the advantages of the technical school, which can only be done by making the preparation more practical in its nature. Take the time allotted for education up to twenty-one, which is our American limit. At sixteen put the boy, no matter what his specialty may be, into a machine shop for a year; teach him to chip and file, and to hand-tool a shaft; to be independent of the labor-saving machines, and also teach him to *work*, and to know that a half dollar represents ten hours' labor. Then at seventeen let him go back to the school. His mental powers have only lain dormant, and a short time will place him on full equality with his comrades, and in advance of them, owing to his practice. His drawings from models will be far more easily and thoroughly acquired. After his three years' study, at twenty, let him take the final year in the shop, learning setting up, fitting of parts, the use of patterns, etc.

Now at twenty-one he is ready to start. Suppose he is to be a blast-furnace man. He goes to the weigh scales and weighs iron. His weights will be accurate, his iron properly graded, and his yard systematically arranged. If his scales get out of order he will adjust them himself. He will gradually gain the confidence of the old founder, who despises "theory blowers," and believes in nothing but practice, but who fully appreciates his honest work and system, and he will absorb information as a sponge does water.

In short, what we want is not so much to improve the technical school as to see that the student shall get all it has to give.

DR. H. WEDDING, *of Germany.*

Our German institutions for technical education have been so often mentioned, that I think it may be perhaps not quite uninteresting to you to hear some more particulars about the technical schools, as far as they are designed for teaching mining and metallurgy.

There are two essentially different kinds of institutions. One kind, which are called Mining Schools (Bergschulen or Steigerschulen), and another, which are called Mining Academies (*Bergakademien*). The first are established in the largest mining districts, and are only designed for the education of working-men; those workmen who show a desire for a higher education, and excel in industry and good behavior, are selected by the owners or the directors, and sent to the mining schools, where they get free lessons, their expenses for living being often paid besides. They have there the opportunity to learn all that is necessary to become good *foremen* (steiger), but it is not designed to make *engineers* out of them.

Sometimes, it is true, one of these young men will show such talent in the school that he will be sent afterwards to a mining academy, but that is only the exception.

The method of education of *engineers*, that is, men who shall have the faculty to manage minės and melting works of every description and size, control the men, superintend the assays, make constructions of machinery, furnaces, etc., is quite in another way. A young man who would become a mining or metallurgical engineer, ought to go through in the first instance a general school (*gymnasium, realschule*), where he gets only a general education, not the least a technical one. When about twenty years old he leaves this school, and goes to the *Bergakademie*, the real technical school. These mining academies are in Germany sometimes institutions for mining branches alone, and sometimes include other branches, as civil engineering, architecture, etc.

We professors of these academies, or rather most of us, try to get up the preliminary general education as high as possible, and prefer the scholars who have a finished general education, finding that these generally make quicker progress than others, though their mathematical and chemical knowledge are very low on the average.

Between the general school and entering of the mining academy, one or two years of practical work in mines or smelting works were formerly prescribed; but this prescription is now omitted, it being free to the scholar to do so still, or come at once to the academy, because we are of the opinion that every young man should be fitted by the general school morally as well as intellectually to take care of himself.

But, as the young men or their parents generally ask the professors for their advice, and this advice is always given for some practical work before entering the academy, most young men go first to a mine, a colliery, or a smelting works for at least one year. For a diligent young man we think one year generally sufficient to get the practical knowledge which is undoubtedly necessary for the understanding of the lectures given at the academy.

The purpose of this practical year is neither to make a workman out of the young scholar, nor an engineer. He cannot in a year become equally a good puddler, coal-cutter, smelter, etc. He can only get a practical idea which will enable him to follow the lectures without obliging the professors to dwell upon practical things in a tedious manner. It is not necessary that the student shall see all different kinds of mines and furnaces. If he has studied a colliery well, he will understand easily the description of a lead mine; if he has studied a blast-furnace well, he will also understand the description of the puddling process, etc.

I always advise the young men to take one branch, for instance puddling, quite earnestly. In working with their own hands, they not only learn, but appreciate, the powers of a laborer; they know what they can demand from a man. I speak here from my own experience. After I finished the *gymnasium*, I went by the advice of our well-known German metallurgist, Mr. Karsten, to a Silesian iron-mill, and learned to work a charcoal finery till I was quite master of this process. I found it afterwards very easy to manage and understand all other iron processes, and I appreciate and value to-day, in my present situation as consulting engineer to the Prussian government for its metallurgical works, the knowledge which I got in those two years of practical work.

The mining academy takes three years. The question arises, is it good to mingle practice with the studies? From my own experience I must question that. The vacations may be well adapted to continue the practical work, or to make excursions, to travel, etc., but study and practical work together seem very irrational.

After the studies at the mining academy are finished, the young man is generally twenty-five years old, and thinks now about securing for himself a situation, which will enable him to be married at thirty. He has now no more time in which to work as a workman. He now begins his career at a mine or a smelting-works; at first working just according to his inclination for mechanical or chemical pursuits, either in a draughting office or in a laboratory, whence he makes his way as an engineer to the place for which his ingenuity and his assiduity may fit him.

So the young man leaves the mining academy not *as* an engineer, but with all the foundation to *become* a good and useful engineer.

www.ingramcontent.com/pod-product-compliance
Lightning Source LLC
LaVergne TN
LVHW021411110826
845150LV00007B/1870

* 9 7 8 1 4 2 5 5 1 0 7 5 6 *